SpringerBriefs in Materials

For further volumes:
http://www.springer.com/series/10111

Zhenyu Li · Ce Wang

One-Dimensional Nanostructures

Electrospinning Technique and Unique Nanofibers

 Springer

Zhenyu Li
Alan G. MacDiarimid Institute
Jilin University
Changchun
People's Republic of China

and

Australian Future Fibres Research
  and Innovation Centre
Institute for Frontier Materials
Deakin University
Geelong
Australia

Ce Wang
Alan G. MacDiarimid Institute
Jilin University
Changchun
People's Republic of China

ISSN 2192-1091   ISSN 2192-1105   (electronic)
ISBN 978-3-642-36426-6   ISBN 978-3-642-36427-3   (eBook)
DOI 10.1007/978-3-642-36427-3
Springer Heidelberg New York Dordrecht London

Library of Congress Control Number: 2013933293

Printed on acid-free paper

Springer is part of Springer Science+Business Media (www.springer.com)

Preface

One-dimensional (1D) nanostructures (wires, rods, tubes, fibers, and belts) are of current interest for their applications in mesoscopic physics and nanoscale devices. In contrast to other nanostructures, 1D nanostructures can provide unique advantages for the investigation of the dependence of electrical, thermal, and mechanical performances on dimensionality. To apply the 1D nanostructures into existed macroscopic devices, there is a great need to develop a novel synthesis route for 1D nanostructures with nanoscaled diameter and macroscopic length. Since the 1990s, a novel and simple technique entitled electrospinning, revived by Reneker, has attracted numerous attention for the generation of 1D nanostructures with continuous length, tuneable diameter, aligned direction, diverse and controllable compositions. Till date, the electrospinning field has been evaluated as the Fast Moving Front in materials science by Thomson, ISI. We therefore publish a mini book to introduce electrospinning. We nevertheless extend our apologies to those scientists whose research findings could not be cited or discussed in our mini book. The present book shall be of interest to those scientists engaged in 1D nanostructures and their applications.

Zhenyu Li
Ce Wang

Contents

Chapter 1
Introduction of Electrospinning

Abstract In the past two decades, electrospinning, as a novel and simple synthesis method for one-dimensional nanostructures, has attracted immensely attentions. During the electrospinning process, electric field force acts on the polymer solution or melt to form the electrospun jet. Finally, solidified fiber can be achieved by stretching the electrified jet for the electrostatic repulsions between the surface charges and the evaporations of solvent. In this chapter, a brief introduction of electrospinning, the history of electrospinning, and basic electrospinning setups/related theories will be presented in turn.

Keywords One-dimensional nanostructures · Electrospinning · Basic setups · Theories

In Sect. 1.1, we provide the brief introduction of electrospinning. The history and basic electrospinning setups/related theories will be presented in Sects. 1.2 and 1.3.

1.1 Brief Introduction of Electrospinning

Nanostructures, having at least one dimension at nanoscale, have gained huge interesting driven by their unique and fascinating properties over their bulk counterparts [1–6]. Among those existed nanostructures, one-dimensional (1D) nanostructures such as wires, rods, tubes, fibers, and belts are of current interest for their unique applications in mesoscopic physics and nanoscaled devices [6, 7]. In contrast to other nanostructures (e.g., quantum dots, quantum wells), the development of 1D nanostructures was slow till 1990s for the difficulties related to the synthesis of those nanostructures with defined morphologies, purities, chemical compositions. From 1990s, a number of advanced techniques have been developed for the generation of 1D nanostructures including electron-beam or focused-ion-beam writing, lithography, hydrothermal, chemical vapor deposition, electrospinning, solution method, etc. [7]. Among those methods, electrospinning (*a concept used to describe a class of fiber forming processes in which electrostatic forces are*

Z. Li and C. Wang, *One-Dimensional Nanostructures*,
SpringerBriefs in Materials, DOI: 10.1007/978-3-642-36427-3_1,
© The Author(s) 2013

employed to control the production of fibers: revived by Reneker in 1990s), as the simplest approach to fabricate 1D nanostructures with both solid and hollow interiors with continuous length, tunable diameter, aligned direction, diverse, and controllable composition, *attracts* numerous attention [8–10]. Different from other methods for 1D nanostructures, electrospinning is based on the electric field force acting on the polymer solution or melt, which can be regarded as a variant of the electrospray process. In electrospinning, solidified fiber can be achieved by stretching the electrified jet for the electrostatic repulsions between the surface charges and the evaporations of solvent [11]. Electrospinning can be applied to synthetic and natural polymers, polymer alloys, and polymer decorated with functional nanomaterials. Those unique advantages of electrospinning afford the multifunctional properties for diverse applications.

Envied by those unique advantages, a remarkable growth and interest have been engaged in electrospinning field. Till now, almost 9,000 scientific papers on electrospinning have been published and the cited times are more than 130,000 as shown in Fig. 1.1.

Currently, electrospinning field has been evaluated as the fast-moving front in materials science by Thomson, ISI. Till now, the electrospinning technique has become more interdisciplinary, numerous cooperations have been established among the experts in diverse research fields. The major objectives of this book aim to provide personal perspective of some special activities in electrospinning technology.

1.2 History of Electrospinning

The process of electrospinning was firstly patented by J. F. Gooley in 1900 (Cooley, J. F. Patent GB 06385 "Improved methods of and apparatus for electrically separating the relatively volatile liquid component from the component of

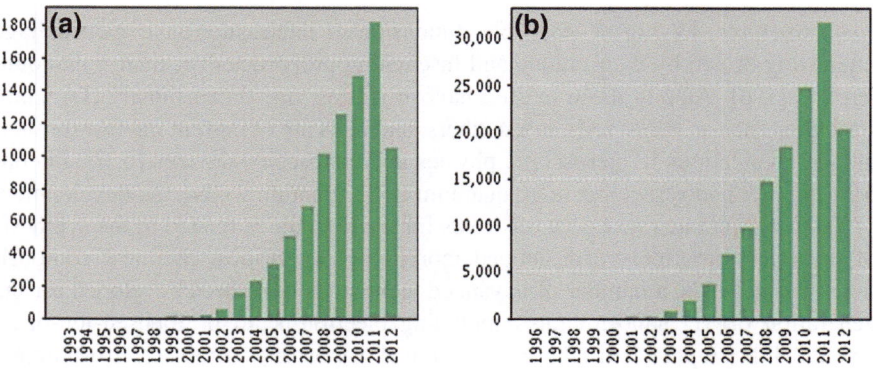

Fig. 1.1 Numbers of scientific publications (**a**) and cited times (**b**) on electrospinning from 1990s with the keyword "electrospinning" or "electrospun" (source web of science). *Date* (31/07/2012)

relatively fixed substances of composite fluids" 19th May 1900). Further development toward commercialization was made by Anton Formals (*U. S. Patent 1,975,504*) in 1934 for the generation of textile yarns. In 1936, C. L. Norton described the patent (*U. S. Patent 2,048,651*) on melt electrospinning. Before the early 1990s, more than 30 US patents on electrospinning have been issued.

Since the early 1990s, Reneker and co-workers (at the University of Akron) revived the electrospinning technique for the generation of one-dimensional (1D) polymer nanostructures, electrospinning technique has been a subject of wide research. Different from the other routes for the synthesis of 1D nanostructure, the formation of solidified electrospun nanostructures (typically on the micro or nano scale) derived from the high extensional deformations and strain rates, caused by polarized charge repulsion originated from the strong polarization under the high electric field, of a viscoelastic jet based on organic solutions or melt. Despite the simplicity of the electrospinning setup, the electrospinning mechanisms are rather complicated including Taylor Cone theory [12], Bending Instability theory [11], Electrically Forced Jet-Stability theory [13, 14], and so on [15, 16]. Those theories not only provide a better understanding of mechanisms corresponding to the electrospinning process, but also endow scientists with the good ability of designing novel setups for the further control of electrospun nanostructures performances. Prompted by those theories, electrospun nanostructures have been extended from single solidified polymer nanofibers to polymer/inorganic hybrid nanofibers, inorganic nanofibers, hollow inorganic nanotubes, etc. Additionally, by modification of the electrospun setups, individual nanofiber, aligned nanofibers, and patterned nanofibers have been also achieved. Taking the unique advantages (*extremely long length, high surface area, complex and diverse compositions and structures, and so on*) of electrospun nanostructures, electrospun nanostructures have gained broad applications in tissue engineering, drug delivery, optoelectronics, chemiresistors, catalysis, filters, fiber reinforcement, wounding healing, photoelectronics, FET, magnetic devices, etc.

1.3 Basic Electrospinning Setups and Related Theories

1.3.1 Basic Electrospinning Setup

Basic electrospinning setup contains three major components: a DC or AC high-voltage power supply, a metallic or plastic syringe, and a collector. The whole electrospinning setup can be found in Fig. 1.2. The syringe is to host the polymer solution. As high voltage was applied, the drop of polymer solution at the tip of syringe will be polarized and the induced charges will distribute on the surface. Under the influence of strong electrostatic field, the charged polymer will be accelerated toward the collector. In another word, within the electrospinning process, the role of the electrostatic forces is to supplement or replace the

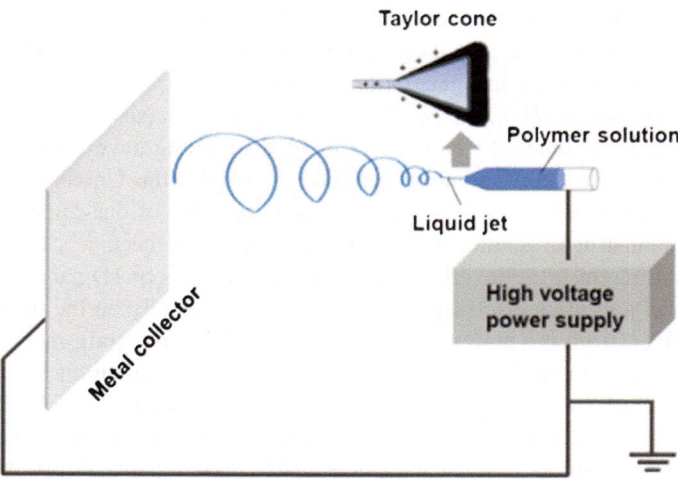

Fig. 1.2   Schematic diagram of the basic electrospinning setup

conventional mechanical forces (e.g., hydrostatic, pneumatic) used to form the jet and to reduce the size of the fiber, hence to term "electrohydrodynamic jetting." The collector, usually, is good electric conductor for the neutralization of the charge carried by nanofibers. However, airborne ions from the corona discharges can also neutralize the charge on the nanofibers in some degree, resulting in that the nanofibers can be collected on isolated collectors or even on the surface of liquid.

1.3.2 Basic Phenomena in Electrospinning and Their Corresponding Theories

Currently, theories on electrospinning can be divided into three parts by observing the phenomena during the process of electrospinning: (1) Theory on the fluid charging; (2) Theories on the liquid droplet under high voltage—Taylor Cone theory; (3) Theories on the jet in flight—Instability Theories.

1.3.2.1 Fluid Charging

In electrospinning, generation of charges on the fluid within the syringe usually occurs by virtue of high electric field polarization between the positive and negative potentials, referred to as induction charging [17]. At this time, free electrons, ions or ion pairs will be generated as charge carriers in the fluid and form double layer in the fluid owing to the ion mobility. In the absence of flow, the double layer thickness is determined by the ion mobility within the fluid; In the presence of flow, ions may be convected away from the electrode and the double layer continually replenished. Inductive charging is generally suitable for fluids with the conductivities of the order of 10^{-2} S/m.

1.3.2.2 Liquid Droplet Under High Voltage—Taylor Cone theory

In 1964, G. I. Taylor established the Taylor Cone theory [16] to describe the deformation of small-volume liquid under the high electric field:

1. As a small volume of electrically conductive liquid is exposed to an electric field, stable shape can be acquired owing to the equilibrium of the electric forces and the surface tension in the cases of inviscid, Newtonian, and viscoelastic liquids (*if the potential is not too large*).
2. As the voltage is increased to the critical potential and any further increase will destroy the equilibrium, thus the liquid body acquires a conical shape, with a half angle of 49.3° (a whole angle of 98.6°), referred to as the **Taylor Cone** as shown in Fig. 1.3.

Additionally, G. I. Taylor also demonstrated that the shape of such a cone approached the theoretical shape just before jet formation within the electrospinning (electrospraying) process. Taylor's derivation is based on two assumptions:

1. The surface of the cone is an equipotential surface
2. The cone exists in steady state equilibrium; namely,

$$U_c^2 = 4\frac{H^2}{L^2}\left(\ln\frac{2L}{R} - \frac{3}{2}\right)(0.117\pi\gamma R) \tag{1.1}$$

where U_c is voltage, H is the distance between the tip of syringe and the collector, L is the length of syringe, R is the diameter of the tip of syringe, and γ is the surface tension of solution.

In 2001, A. L. Yarin and D. H. Reneker modified the Taylor Cone theory based on the experimental data [12]. They claimed that the Taylor Cone corresponded essentially to a specific self-similar solution, whereas there existed nonself-similar

Fig. 1.3 Schematic diagram of the Taylor Cone

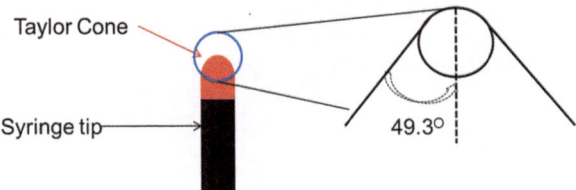

solutions which did not tend toward a Taylor Cone. Namely, there existed another shape, which was not self-similar.

They found that the assumption (1) within the previous Taylor Cone theory was uncorrected. The theory predicted that, as the potential increases and approaches the critical (Rayleigh) value, the stable shape becomes less and less prolate, tending toward a sphere. However, in the experiment an increase in potential resulted in more prolate droplets. Moreover, they also proved that as a liquid surface develops a critical shape, its configuration approaches the shape of a cone with a half angle of 33.5°, rather than the previous angle of 49.3° as shown in Fig. 1.4.

1.3.2.3 Jets in Flight-Thinning and Instability Theories

Scientists in electrospinning field observed that as the charged jets ejected from the Taylor Cone, during the process of electrospinning, the jets usually passed by in a nearly straight line preceded by bending into a complex path, during which electrical forces stretched and thinned them by large ratios (Fig. 1.5). According to those phenomena, lots of theoretical works have been carried on.

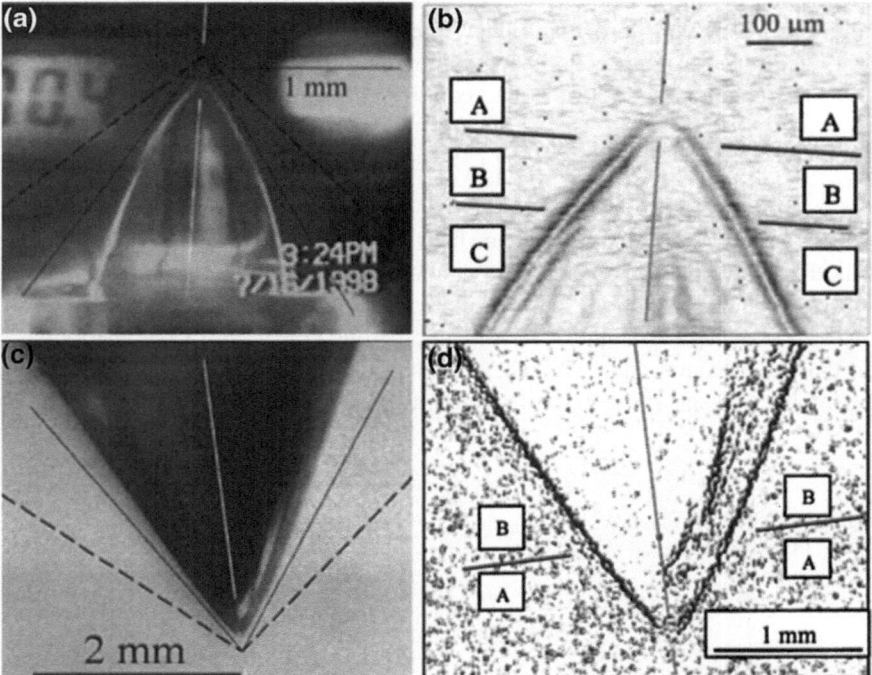

Fig. 1.4 a Videograph of the critical droplet shape observed for a sessile droplet. **b** Part of the image in (**a**), processed with Scion Image. **c** Critical droplet shape observed for a pendant drop. **d** The enlarged droplet tip from (**c**), processed with Scion Image. Reprinted with permission from Ref. [12]. Copyright 2001, American Institute of Physics

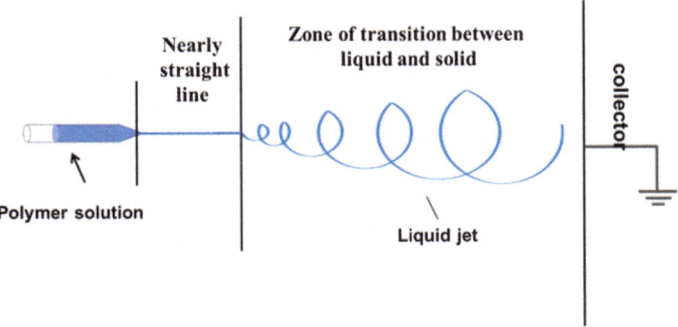

Fig. 1.5 Schematic illustration of the jets path between the tip of syringe and collector

Slender Thinning Jet

As the fluid ejected from the cone, a continuous thinning liquid jet can be observed. The charged jet will accelerate in the presence of the electric field created by the countered positive and negative potential as shown in Fig. 1.6. Numerous attempts have been made to explain such trends in electrospun jet operation as functions of fluid and operating parameters under experimental control. Within certain limits, for a given fluid, there exist ranges of driving voltage (V) and flow rate over which the electrospinning can be maintained stably for long periods. The electrospinning process constitutes an electrical circuit, so one can measure as well the current (I) flowing through this "circuit." Numerous

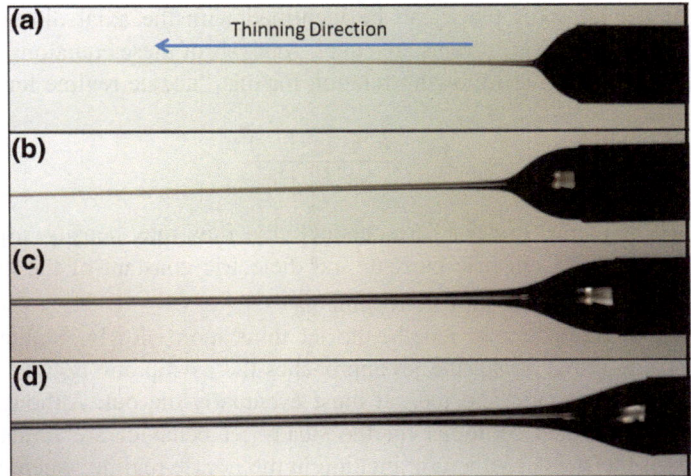

Fig. 1.6 Cone-jets of 2 % solutions of polyethylene oxide (MW = 920 kg/mol) in water. $D = 45$ cm in all cases. **a** $Q = 0.02$ ml/min, $E = 0.282$ kV/cm. **b** $Q = 0.10$ ml/min, $E = 0.344$ kV/cm. **c** $Q = 0.50$ ml/min, $E = 0.533$ kV/cm. **d** $Q = 1.00$ ml/min, $E = 0.716$ kV/cm. Reprinted from Ref. [17], Copyright 2007, with permission from Elsevier

investigators have reported current I increasing with voltage V in a nonlinear fashion. For example, Demir and Erman demonstrated that for sufficiently large flow rates, current I is roughly proportional to flow rate Q, indicative of a volumetric charge density, I/Q, independent of flow rate. At low flow rates, I/Q tends to increase as I becomes less sensitive to Q [18]. Additionally, they also fit their data to power laws, $I - V^x$, with values of $x = 2.17$ and 2.7, respectively. Theron et al. [19] on the other hand, reported $I - Q^x$, $-1.04 < x < 0.26$, for a range of fluids, which they attributed to ion mobility–limited charging of the fluid in their equipment configuration.

Mathematical description of the thinning liquid jet has been formulated within the context of conventional electrohydrodynamics by Gañán-Calvo [20]. In 2001, Hohman et al. [13] established formulated the slender body model for Newtonian jets. This model was shown to describe well the jet shape for low-conductivity fluids. As for high-conductivity fluids, convergence of the equations to a solution was found to be laborious, a problem associated with specifying the unknown initial surface charge density on the fluid jet exiting the syringe. Alternatively, Hartman and co-workers proposed a different boundary condition with specified initial slope $(dh/dz)_0$ at the nozzle [14]. In 2002, Feng further reformulated the slender body model for Newtonian jets using an approximation for the electric field equation in which the convergence problem of Hohman and co-workers has been circumvented [21]. In brief, Hohman and Feng's results indicate that the shape of the jet is strongly dependent on the evolution of surface charge density and local electric field. Concurrent with the initial rapid reduction in jet diameter, the latter two quantities rise quickly to maximum values as charges relax to the jet surface and surface advection current becomes more important relative to bulk conduction current. The characteristic length over which the initial dramatic thinning of the jet takes place can be identified with the axial distance where advection and conduction currents are equal. Analysis of these equations (Fridrikh, unpublished) leads to the following relation for this "nozzle regime length" L:

$$L^5 = \frac{K^4 Q^7 \rho^3 (\ln \chi)^2}{8 \pi^2 E_\infty I^5 \bar{\varepsilon}^2} \tag{1.2}$$

where K, Q, ρ, χ, E_∞, I, and $\bar{\varepsilon}^2$ are conductivity, flow rate, density, aspect ratio ($\chi = \frac{D}{h_0}$), applied field, electric current, and dielectric constant of the outer fluid (typically, air in conventional electrospinning).

Beyond this characteristic length, the jet thins more slowly. Sufficiently far from the nozzle (circa $30h_0$), the jet approaches the asymptotic regime where all terms except electrostatic and inertial must eventually die out. Although illuminating, these analytical solutions for the steady jet behavior are limited by the breakdown of the slender jet approximation in the nozzle regime, where the cone-jet occurs. Numerical methods such as that employed by Yan et al. [22] for the cone-jet can be matched to the slender thinning jet solutions.

Jet Instability

Frankly speaking, the cone-jet operation is sufficient to draw out continuous fibers
to very small diameter. However, the fluids typically used in electrospinning do
not always solidify sufficiently en route to the collector to remain fibrous after
impact on the collector. In practice, as the jet thins, it ultimately succumbs to one
or more fluid instabilities which deform the jet as they grow. A family of such
instabilities exists and can be analyzed for different conditions of the symmetry
(axisymmetric or nonaxisymmetric) of the growing perturbations of the jet, using
linear instability analysis. Figure 1.7 illustrates the perturbations associated with
several of the lowest-order instabilities.

Although splaying of the jet was proposed in the 1990s as the primary mode of
instability [8], and some evidence has been found for splitting in "postmortem"

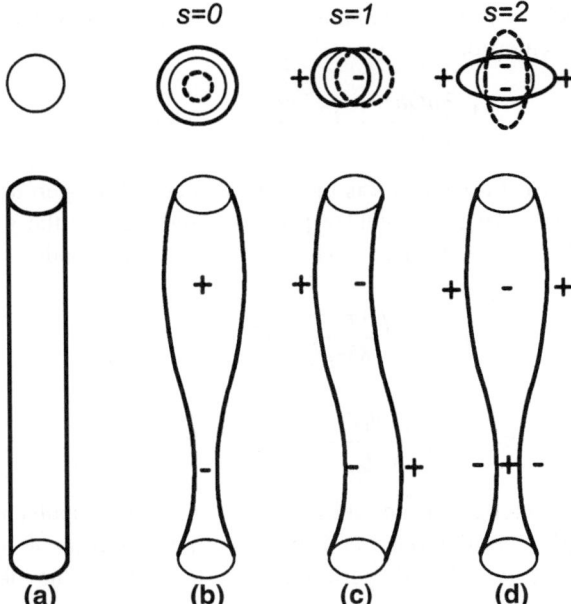

Fig. 1.7   Schematic illustration of perturbations associated with several of the lowest-order
instabilities, distinguished by their azimuthal wave number, s. Top views illustrate cross sections
of the jet at maximum amplitudes of (oscillatory) perturbation, with bold and dashed contours
representing different positions along the jet length. Bottom views illustrate changes in shape and
center line down the length of the jet. $\pm$ are used to indicate regions of positive or negative
deviation from the unperturbed jet shape. Perturbations are exaggerated beyond the linear
instability regime **a** unperturbed cylindrical fluid element, **b** varicose ($s = 0$) instability,
c whipping ($s = 1$) instability (also called "bending" or "kink" instability in the literature), and
d splitting ($s = 2$) instability. Growth of the varicose instability leads to equal-sized droplets;
growth of the splitting instability leads to two equal-sized sub-jets. Higher-order ($s > 2$)
instabilities are also conceivable. Reprinted from Ref. [17], Copyright 2007, with permission
from Elsevier

micrographs of fibers [23] and for secondary jetting in images of jets [11], such events are relatively rare; the most common mode of instability in electrospun jets appears to be the growth of lateral excursions of the jet, the so-called whipping model [11, 16, 24], this instability is not a consequence of viscoelasticity of the fluid and may occur for Newtonian and non-Newtonian fluids as well. The main competing mode of instability, for the relatively dilute solutions often employed to achieve the smallest fiber diameters in electrospinning, is that of droplet breakup (electrospraying). For the steady thinning jets described above, Hohman et al. [13] analyzed the linear instability analysis for both droplet breakup and whipping modes and produced "operating diagrams" for electrospinning that illustrate the combinations of controllable parameters (flow rate, Q, and applied field, $E_\infty = V/D$) under which a particular fluid will electrospin (whipping) rather than electrospray (droplet breakup). Upon whipping, charge repulsion is relatively stronger in contrast to electrical shear stress and surface tension, allowing one to neglect the contributions to the dispersion relation arising from the electric field, surface tension, and the finite conductivity. Within this approximation, considerable simplification of the dispersion relation is achieved.

$$-\omega k^4 \frac{3\rho Q h^2}{4\pi^3 \mathrm{Re} h_0} + k^2 \frac{2\pi\sigma^2}{\bar{\varepsilon}}(3 - 2\ln(\chi)) = \omega^2 \rho \tag{1.3}$$

where ω is the growth rate and k is the wave number of the perturbation. Solving for ω and differentiating this with respect to k, one obtains the growth rate and wave number for the most unstable (i.e., fastest growing) mode:

$$\omega_{\max} = \left(\frac{9\pi^5 \sigma h_{0\mathrm{Re}}}{8h^2 \rho^2 Q_{\bar{\varepsilon}}^2}(2\ln(\chi) - 3)^2\right)^{\frac{1}{3}} \tag{1.4}$$

$$k_{\max} = \left(\frac{\pi^3 \sigma h_0 \mathrm{Re}}{h^2 Q_\rho}\right)^{\frac{1}{3}}\left(\frac{2\pi\rho}{\bar{\varepsilon}}(2\ln(\chi) - 3)\right)^{\frac{1}{6}} \tag{1.5}$$

These equations reflect the destabilizing effect of charge repulsion (strongest for short-wavelength perturbations) and the stabilizing effect of viscosity (also strongest for short-wavelength perturbations), as well as the damping of the instability growth rate due to inertia.

It should be note that multiple instabilities can also grow simultaneously, and even a whipping jet can ultimately decay into droplets (Fig. 1.8).

Hohman et al. also reported a second, electrically driven droplet breakup instability that becomes important at high electric fields in fluids of finite conductivity, in addition to the conventional, surface tension–driven instability. They also obtained an expression for the lateral growth of the jet excursions arising from the whipping instability far from its onset, deep in the nonlinear regime. This equation contains terms due to acceleration of the charged jet under the influence of the applied electric field, normal stresses due to surface tension and bending of electric field lines, and charge–charge repulsion. As the diameter of the jet

Fig. 1.8 Whipping jet decay into droplets. Reprinted from Ref. [17], Copyright 2007, with permission from Elsevier

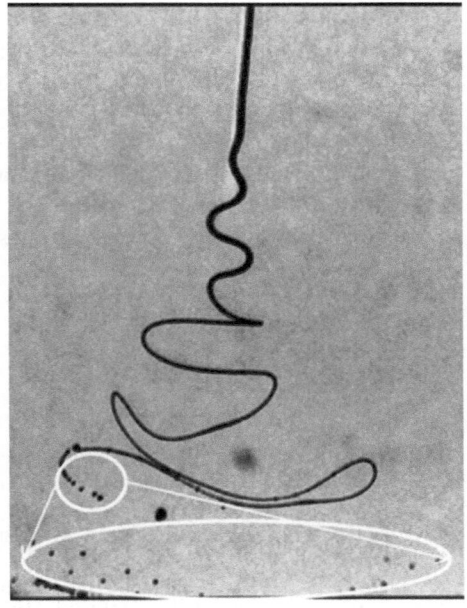

becomes small due to the growth of the whipping instability, the longest-lived terms are those due to surface tension and charge repulsion. The competition between surface tension and charge repulsion can be used to define a "stability parameter" S:

$$\frac{1}{S} = \left(\frac{\pi(2 \ln \chi - 3)}{2\gamma\bar{\varepsilon}}\right)^{\frac{1}{3}} \left(\frac{I}{Q}\right)^{\frac{2}{3}} h \tag{1.6}$$

As $S < 1$: instability with respect to the whipping mode. However, rapidly increases to 1 as h is reduced under the action of the growing lateral perturbations.

As $S = 1$: the destabilizing effect of charge repulsion is exactly balanced by the stabilizing effect of surface tension, leading to the cessation of stretching and definition of a limiting "terminal jet" diameter.

It can be found that this is only a limiting diameter arising from a balance of forces on the Newtonian fluid; elastic forces arising from solution viscoelasticity or solvent evaporation may intervene under certain condition to arrest the jet prematurely with large diameter. While a similar analysis of the elongation of the steady jet in a tangential electric field indicates that very small diameter jets can be realized in this case, too, before surface tension forces eventually rise to balance the electric shear stress.

As illustrated in Fig. 1.8, jets will break into drops without or very low elasticity. Elastic effects thus can arise as a consequence of the viscoelasticity of the fluid itself or development of a solid-like elasticity as the jet cools or solvent evaporates en route to the collector. It is usually postulated that the growth of the whipping instability is responsible for large stretching of the fluid jet and lower the jet diameter, which can provide greater opportunity for solvent evaporation and jet solidification in the whipping regime. Stephens and Bruce demonstrated that the solvent evaporation prior to the onset of instability could be negligible through Raman spectrum [25].

Although details of the theoretical analysis have not illustrated in this book, they do construct a powerful platform to better understand the mechanisms corresponding to the electrospinning process and design new setups for functional nanofibers.

References

1. Bein T, Stucky GD (1996) Preface to the special issue. Chem Mater 8(8):1569–1570
2. Alivisatos P et al (1998) From molecules to materials: current trends and future directions. Adv Mater 10(16):1297–1336
3. Ozin GA (1992) Nanochemistry: synthesis in diminishing dimensions. Adv Mater 4(10):612–649
4. Schulz W (2000) Nanotechnology: the next big thing. Chem Eng News Arch 78(18):41–47
5. Thiaville A, Miltat J (1999) Small is beautiful. Science 284(5422):1939–1940
6. Xia Y et al (2003) One-dimensional nanostructures: synthesis, characterization, and applications. Adv Mater 15(5):353–389
7. Wang ZL (2000) Characterizing the structure and properties of individual wire-like nanoentities. Adv Mater 12(17):1295–1298
8. Doshi J, Reneker DH (1995) Electrospinning process and applications of electrospun fibers. J Electrostat 35(2–3):151–160
9. Reneker DH, Chun I (1996) Nanometre diameter fibres of polymer, produced by electrospinning. Nanotechnology 7(3):216–223
10. Srinivasan G, Reneker DH (1995) Structure and morphology of small diameter electrospun aramid fibers. Polym Int 36(2):195–201
11. Reneker DH et al (2000) Bending instability of electrically charged liquid jets of polymer solutions in electrospinning. J Appl Phys 87(9):4531–4547
12. Yarin AL, Koombhongse S, Reneker DH (2001) Taylor Cone and jetting from liquid droplets in electrospinning of nanofibers. J Appl Phys 90(9):4836–4846
13. Hohman MM et al (2001) Electrospinning and electrically forced jets. I. Stability theory. Phys Fluids 13(8):2201–2220
14. Hohman MM et al (2001) Electrospinning and electrically forced jets. II. Applications. Phys Fluids 13(8):2221–2236
15. Feng JJ (2003) Stretching of a straight electrically charged viscoelastic jet. J Nonnewton Fluid Mech 116(1):55–70
16. Shin YM et al (2001) Experimental characterization of electrospinning: the electrically forced jet and instabilities. Polymer 42(25):09955–09967
17. Rutledge GC, Fridrikh SV (2007) Formation of fibers by electrospinning. Adv Drug Delivery Rev 59(14):1384–1391
18. Demir MM et al (2002) Electrospinning of polyurethane fibers. Polymer 43(11):3303–3309
19. Theron SA, Zussman E, Yarin AL (2004) Experimental investigation of the governing parameters in the electrospinning of polymer solutions. Polymer 45(6):2017–2030
20. Gañán-Calvo AM (1997) Cone-Jet analytical extension of Taylor's electrostatic solution and the asymptotic Universal scaling laws in electrospraying. Phys Rev Lett 79(2):217–220

21. Feng JJ (2002) The stretching of an electrified non-Newtonian jet: a model for electrospinning. Phys Fluids 14(11):3912–3926
22. Yan F, Farouk B, Ko F (2003) Numerical modeling of an electrostatically driven liquid meniscus in the cone–jet mode. J Aerosol Sci 34(1):99–116
23. Bergshoef MM, Vancso GJ (1999) Transparent nanocomposites with ultrathin, electrospun nylon-4, 6 fiber reinforcement. Adv Mater 11(16):1362–1365
24. Baumgarten PK (1971) Electrostatic spinning of acrylic microfibers. J Colloid Interface Sci 36(1):71–79
25. Stephens JS et al (2001) "Real time" Raman studies of electrospun fibers. Appl Spectrosc 55(10):1287–1290

Chapter 2
Effects of Working Parameters on Electrospinning

Abstract Working parameters are very important to understand not only the nature of electrospinning but also the conversion of polymer solutions into nanofibers through electrospinning. Those parameters can be broadly divided into three parts such as solution parameters, process parameters, and ambient parameters. Each of those parameters can affect the fibers morphologies and by proper control of those parameters we can fabricate electrospun fibers with desired morphologies and diameters. In this chapter, we will present a concise introduction of those parameters and their influence on fiber properties.

Keywords Electrospinning · Solution parameters · Process parameters · Ambient parameters

In Sect. 2.1, we describe the affections of solution parameters on the conversion of polymer solutions into nanofibers during the electrospinning. The affections of process parameters and ambient parameters will be discussed in Sects. 2.2 and 2.3. For each of these affections, examples are presented, respectively.

2.1 Solution Parameters

2.1.1 Concentration

The concentrations of polymer solution play an important role in the fiber formation during the electrospinning process. Four critical concentrations from low to high should be noted:

1. As the concentration is very low, polymeric micro (nano)-particles will be obtained. At this time, electrospray occurs instead of electrospinning owing to the low viscosity and high surface tensions of the solution [1].
2. As the concentration is little higher, a mixture of beads and fibers will be obtained [2–4].

Z. Li and C. Wang, *One-Dimensional Nanostructures*,
SpringerBriefs in Materials, DOI: 10.1007/978-3-642-36427-3_2,
© The Author(s) 2013

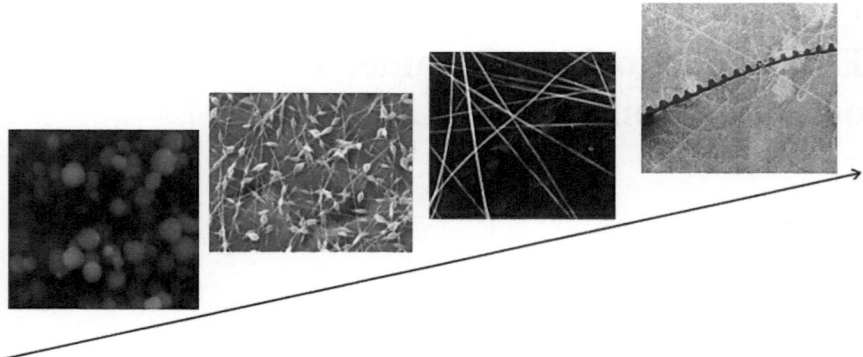

Fig. 2.1 SEM images of the evolution of the products with different concentrations from low to high during the electrospinning. Reprinted from Ref. [3], copyright 1999, with permission from Elsevier, reproduced from Ref. [5] by permission of John Wiley & Sons Ltd

3. When the concentration is suitable, smooth nanofibers can be obtained [2–4].
4. If the concentration is very high, not nanoscaled fibers, helix-shaped micro-ribbons will be observed [5].

To clearly see the evolution of the products with different critical concentrations from low to high, four typical SEM images have been used to illustrate the whole change (Fig. 2.1).

Usually, increasing the concentration of solution, the fiber diameter will increase if the solution concentration is suitable for electrospinning. Additionally, solution viscosity can be also tuned by adjusting the solution concentration.

2.1.2 Molecular Weight

Molecular weight of the polymer also has an important effect on morphologies of electrospun fiber. In principle, molecular weight reflects the entanglement of polymer chains in solutions, namely the solution viscosity. Keep the concentration fixed, lowering the molecular weight of the polymer trends to form beads rather than smooth fiber. Increasing the molecular weight, smooth fiber will be obtained. Further increasing the molecular weight, micro-ribbon will be obtained (Fig. 2.2) [6].

It is also important to note that too high molecular weight favors the formation of micro-ribbon even with the low concentration as shown in Fig. 2.3 [7]. Additionally, the authors also found that as the molecular weight is very high, some patterned fibers can also be obtained at low concentration (Fig. 2.4).

However, it is also important to point out that the molecular weight is not always essential for electrospinning if sufficient intermolecular interactions can be supplied by oligomers. For example, Long and McKee successfully obtained the

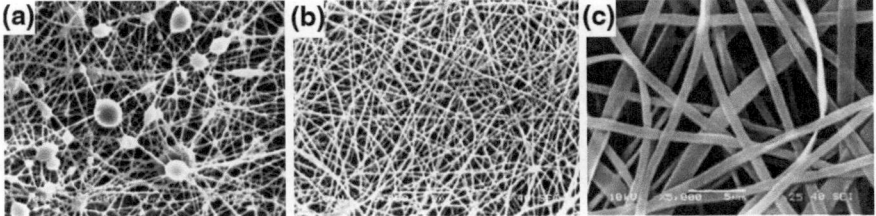

Fig. 2.2 Photographs showing the typical structure in the electrospun polymer for various molecular weights. **a** 9000–10,000 g/mol; **b** 13,000–23,000 g/mol; and **c** 31,000–50,000 g/mol (solution concentration: 25 wt. %). Reprinted from Ref. [6], copyright 2004, with permission from Elsevier

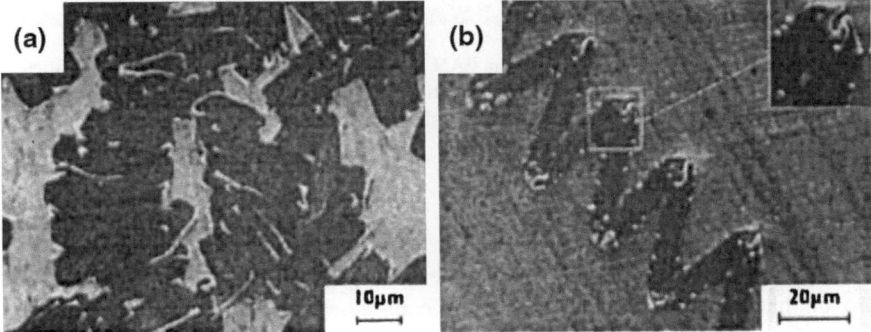

Fig. 2.3 SEM images of the morphologies of the polyacrylamide with ultrahigh MW (9×10^6 g/mol). The concentration of **a** and **b** is 2 wt. % and 2.5 wt. %, respectively. Reproduced from Ref. [7] by permission of John Wiley & Sons Ltd

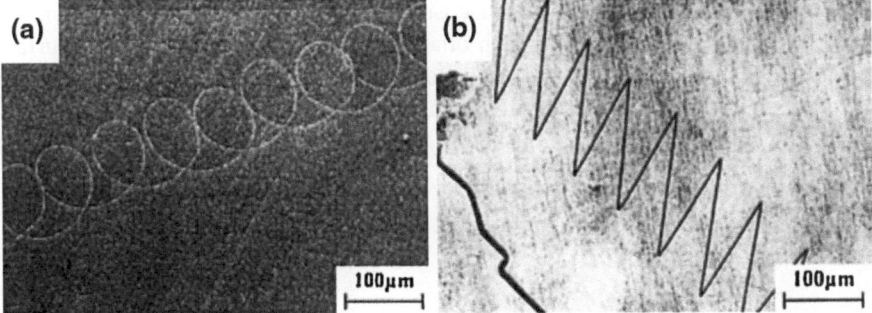

Fig. 2.4 SEM images of the helical fibers and zigzag ribbon. The molecular weight of the polyacrylamide is 9×10^6 g/mol. Reproduced from Ref. [7] by permission of John Wiley & Sons Ltd

oligomer-sized phospholipid fibers from lecithin solutions via electrospinning. They found that as the concentration of phospholipid is higher than 35 wt. %, smooth fiber could be obtained [8].

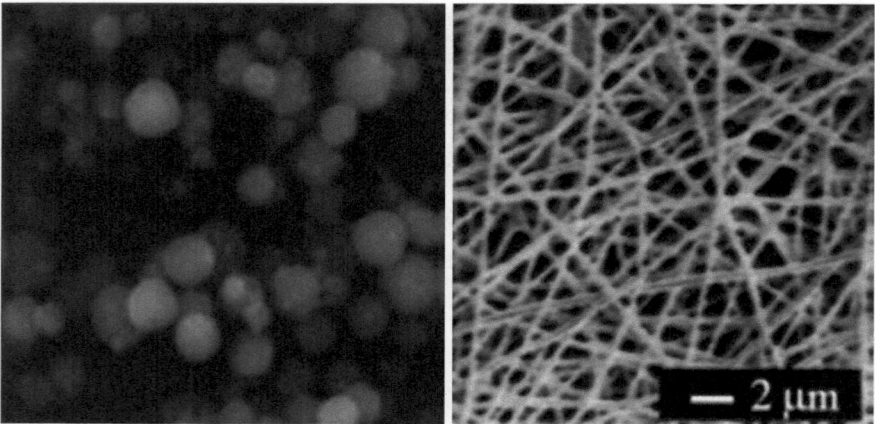

Fig. 2.5 SEM images of the electrospun PAN products with different solution viscosities by adjusting the concentration of the polymer solution. The concentrations of left and right are 1.3 and 15 wt. %, respectively. The molecular weight of PAN is 150,000

2.1.3 Viscosity

Solution viscosity is the critical key in determining the fiber morphology. It has been proven that continuous and smooth fibers cannot be obtained in very low viscosity, whereas very high viscosity results in the hard ejection of jets from solution, namely there is a requirement of suitable viscosity for electrospinning [9, 10]. Generally, the solution viscosity can be tuned by adjusting the polymer concentration of the solution; thus, different products can be obtained as shown in Fig. 2.5. The viscosity range of a different polymer or oligomer solution at which electrospinning is done is different. It is important to note that viscosity, polymer concentration, and polymeric molecular weight are related to each other. For solution of low viscosity, surface tension is the dominant factor and just beads or beaded fiber formed. If the solution is of suitable viscosity, continuous fibers can be obtained. A number of papers on such interrelationships have been published [6, 11–15].

2.1.4 Surface Tension

Surface tension, as the function of solvent compositions of the solution, is quite important factor in electrospinning. In 2004, Yang and Wang systematically investigated the influence of surface tensions on the morphologies of electrospun products with PVP as model with ethanol, DMF, and MC as solvents (Fig. 2.6) [5]. They found that different solvents may contribute different surface tensions. With the concentration fixed, reducing the surface tension of the solution, beaded fibers can be converted into smooth fibers.

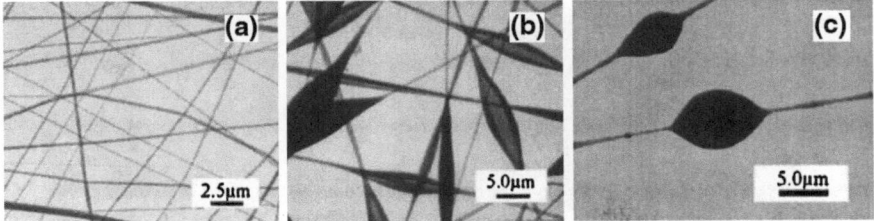

Fig. 2.6 TEM images of the PVP nanofibers electrospun from ethanol (**a**), MC (**b**), and DMF (**c**), respectively. The concentration of PVP is fixed at 4 wt. %. Reproduced from Ref. [5] by permission of John Wiley & Sons Ltd

Additionally, they also demonstrated that the surface tension and solution viscosity can been adjusted by changing the mass ratio of solvents mix (Fig. 2.7) and fiber morphologies (Fig. 2.8).

Basically, surface tension determines the upper and lower boundaries of the electrospinning window if all other conditions are fixed [16–18].

2.1.5 Conductivity/Surface Charge Density

Solution conductivity is mainly determined by the polymer type, solvent sort, and the salt. Usually, natural polymers are generally polyelectrolytic in nature, in which the ions increase the charge carrying ability of the polymer jet, subjecting to higher tension under the electric field, resulting in the poor fiber formation in contrast to the synthetic counterpart [19]. Additionally, the electrical conductivity of the solution can be tuned by adding the ionic salts like KH_2PO_4, NaCl, and so on [20]. With the aid of ionic salts, nanofibers with small diameter can be obtained. Sometimes high solution conductivity can be also achieved by using organic acid as the solvent. For example, Hou et al. [20] used formic acid as the solvent to

Fig. 2.7 Surface tension and viscosity as functions of mass ratio of ethanol/DMF. Reproduced from Ref. [5] by permission of John Wiley & Sons Ltd

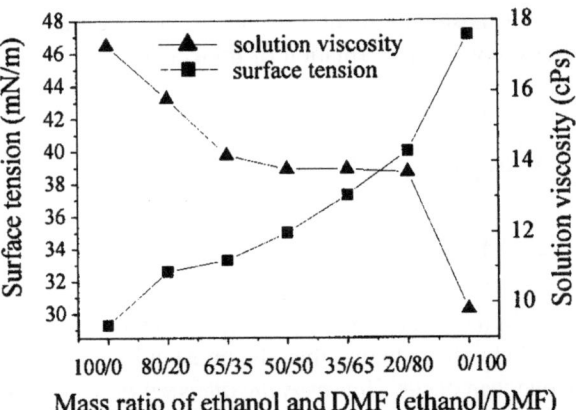

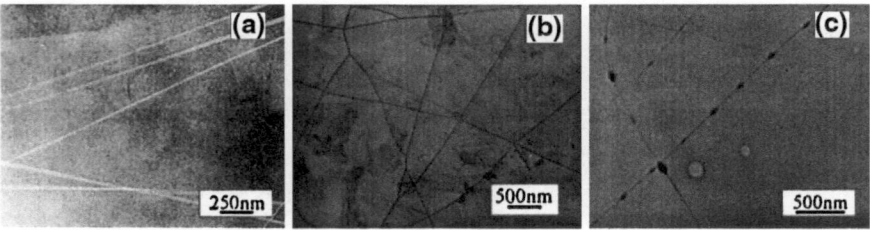

Fig. 2.8 TEM images of PVP (4 wt. %) nanofibers electrospun from ethanol/DMF solution with different mass ratios: **a** 65/35, **b** 50/50, and **c** 35/65, respectively. Reproduced from Ref. [5] by permission of John Wiley & Sons Ltd

Fig. 2.9 SEM images of the electrospun products from 2 wt. % nylon-4, 6/formic acid solution. *Left* is beaded nanofibers. *Right* is bead-free nanofiber by adding 0.44 % pyridine. Reprinted from Ref. [20], copyright 2006, with permission from IOP Publishing Ltd

dissolve the nylon and obtained ultrathin (3 nm) electrospun nylon-beaded fibers (Fig. 2.9). In their study, small amount of pyridine has been also added into the solution to eliminate the beads by increasing the conductivity of the solution.

All those data mentioned above indicate that increase in the solution conductivity favors the formation of thinner fibers.

2.2 Processing Parameters

2.2.1 Voltage

Within the electrospinning process, applied voltage is the crucial factor. Only the applied voltage higher than the threshold voltage, charged jets ejected from Taylor Cone, can occur. *However, the affection of the applied voltages on the diameter of electrospun fibers is a little controversial.* For example, Reneker and Chun [21] have demonstrated that there is not much effect of electric field on the diameter of electrospun polyethylene oxide (PEO) nanofibers. Several groups suggested that higher voltages facilitated the formation of large diameter fiber. For example, Zhang et al. [18] investigated the effect of voltage on morphologies and fiber diameters distribution with poly (vinyl alcohol) (PVA)/water solution as model (Fig. 2.10).

Several groups suggested that higher voltages can increase the electrostatic repulsive force on the charged jet, favoring the narrowing of fiber diameter. For example, Yuan et al. [22] investigated the effect voltage on morphologies and fiber alignment with polysulfone (PSF)/DMAC/acetone as model (Fig. 2.11).

In addition to those phenomena, some groups also demonstrated that higher voltage offers the greater probability of beads formation [1, 23, 24].

Thus, we can found that voltage does influence fiber diameter, but the level of significances varies with the polymer solution concentration and on the distance between the tip and the collector [25].

2.2.2 Flow Rate

The flow rate of the polymer solution within the syringe is another important process parameter. Generally, lower flow rate is more recommended as the polymer solution will get enough time for polarization. If the flow rate is very high, bead fibers with thick diameter will form rather than the smooth fiber with thin diameter owing to the short drying time prior to reaching the collector and low stretching forces. For example, Yuan et al. [23] investigated the effect of the flow rate on the morphologies of the PSF fibers from 20 % PSF/DMAC solution at 10 kV (Fig. 2.12). In their study, bead fibers with thicker diameters can be obtained as the flow rate is 0.66 ml/h.

2.2.3 Collectors

During the electrospinning process, collectors usually acted as the conductive substrate to collect the charged fibers. Generally, Al foil is used as a collector but it is difficult to transfer the collected nanofibers to other substrates for various applications. With the need of fibers transferring, diverse collectors have been developed including wire mesh [26], pin [27], grids [28], parallel or gridded bar [29], rotating rods or wheel [29], liquid bath [30], and so on (Fig. 2.13).

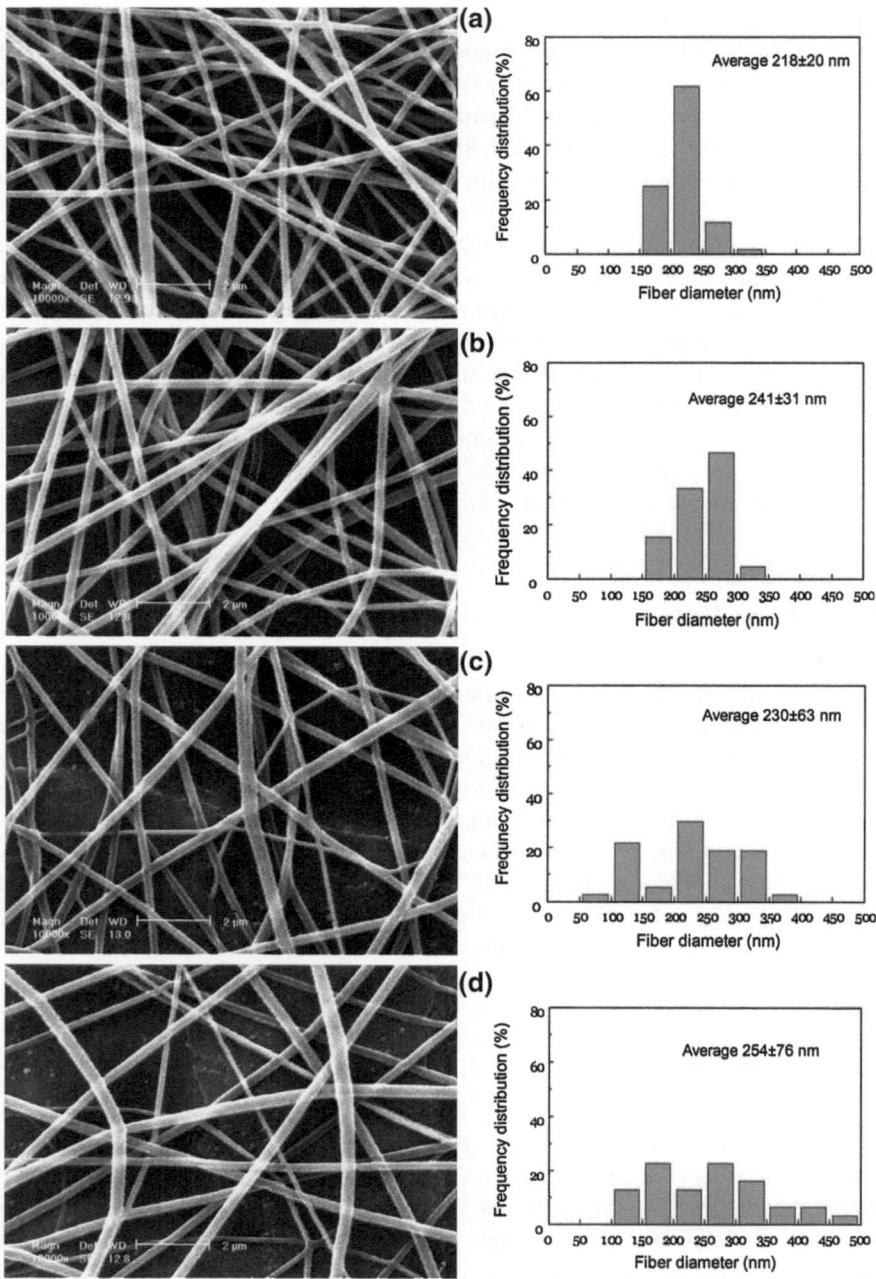

Fig. 2.10 Effect of voltage on morphology and fiber diameter distribution from a 7.4 wt. % PVA/water solution (DH = 98 %, tip–target distance = 15 cm, flow rate = 0.2 ml/h). Voltages: **a** 5; **b** 8; **c** 10; and **d** 13 kV. Reprinted from Ref. [18], copyright 2005, with permission from Elsevier

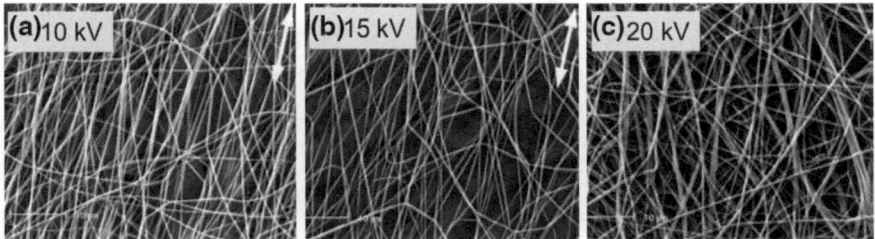

Fig. 2.11 SEM images of the ultrafine fibers electrospun from a 20 % PSF solution in DMAC/ acetone (9:1) under different voltages. The applied voltages of A, B, and C are 10, 15, and 20 kV, respectively. The average diameters of A, B, and C are 344 ± 51, 331 ± 26, and 323 ± 22 nm, respectively. Reproduced from Ref. [22] by permission of John Wiley & Sons Ltd

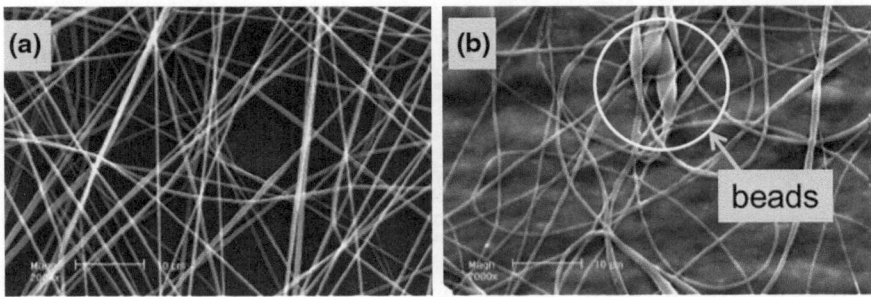

Fig. 2.12 SEM images of the effect of the flow rate on the morphologies of the PSF fibers from 20 % PSF/DMAC solution at 10 kV. Flow rates of A and B are 0.40 and 0.66 ml/h, respectively. Reproduced from Ref. [22] by permission of John Wiley & Sons Ltd

2.2.4 Distance (H) Between the Collector and the Tip of the Syringe

It has been proven that the distance (H) between the collector and the tip of the syringe can also affect the fiber diameter and morphologies [12]. In brief, if the distance is too short, the fiber will not have enough time to solidify before reaching the collector, whereas if the distance is too long, bead fiber can be obtained. It is well known that one important physical aspect of the electrospun fiber is the dryness from the solvent, so optimum distance is recommended. For example, Yuan et al. [22] demonstrated that a little long distance favors the thinner fiber diameter as shown in Fig. 2.14.

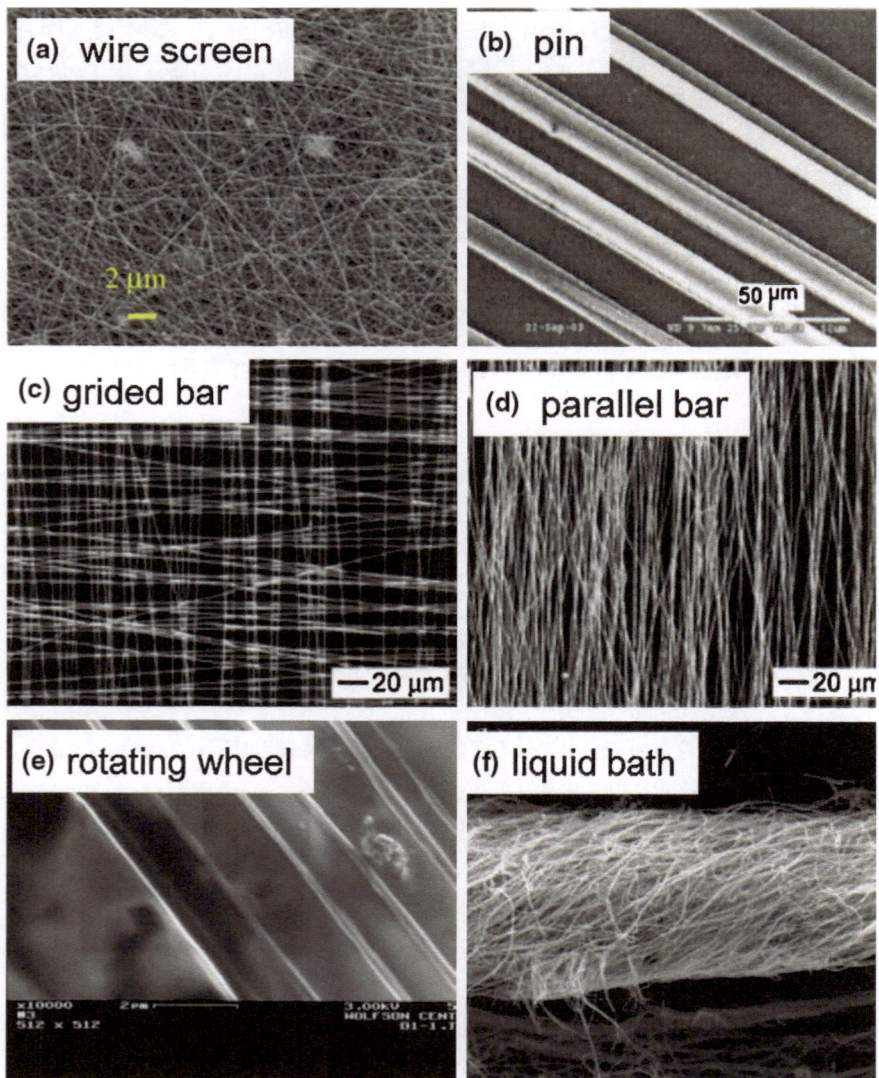

Fig. 2.13 SEM images of the different electrospun products with different types of collectors. **a** Reprinted from Ref. [26], copyright 2005, with permission from Elsevier; **b** Reprinted with permission from Ref. [27]. Copyright 2004, American Institute of Physics; **c** Reproduced from Ref. [28] by permission of John Wiley & Sons Ltd; **d** and **e** Reproduced from Ref. [29] by permission of John Wiley & Sons Ltd; **f** Reproduced from Ref. [30] by permission of John Wiley & Sons Ltd

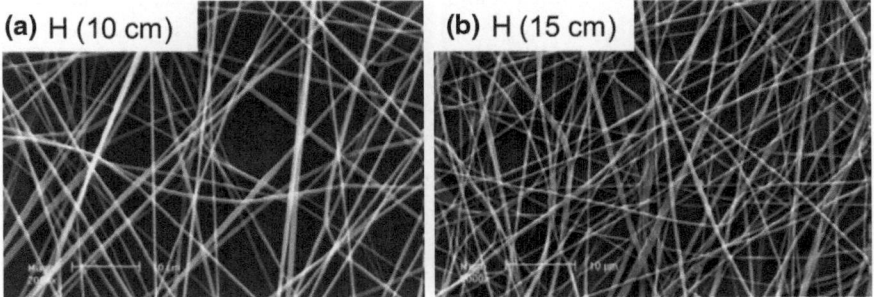

Fig. 2.14 SEM images of the electrospun PSF fibers from 20 wt. % PSF/DMAC solution at 10 kV with different distances. The distances of *A* and *B* are 10 and 15 cm, respectively. The diameters of *A* and *B* are 438 ± 72 and 368 ± 59 nm, respectively. Reproduced from Ref. [22] by permission of John Wiley & Sons Ltd

2.3 Ambient Parameters

Ambient parameters can also affect the fiber diameters and morphologies such as humidity, temperature. For example, Mituppatham et al. [31] had proven that increasing temperature favors the thinner fiber diameter with polyamide-6 fibers for the inverse relationship between the solution viscosity and temperature (as shown in Fig. 2.15).

As for the humidity, low humidity may dry the solvent totally and increase the velocity of the solvent evaporation. On the contrary, high humidity will lead to the thick fiber diameter owing to the charges on the jet can be neutralized and the stretching forces become small. Recently, Casper et al. [32] demonstrated that the variety of humidity can also affect the surface morphologies of electrospun PS fibers (Fig. 2.16).

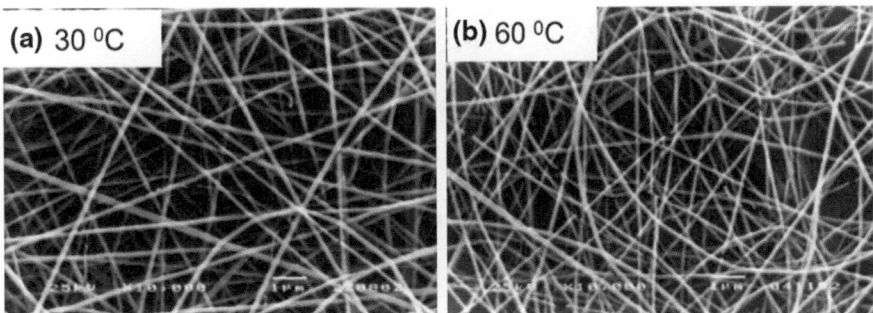

Fig. 2.15 SEM images of the electrospun PA-6-32 fibers under different temperatures. The temperatures of *A* and *B* are 30 and 60 °C, respectively. The diameters of *A* and *B* are 98 and 90 nm, respectively. Reproduced from Ref. [30] by permission of John Wiley & Sons Ltd

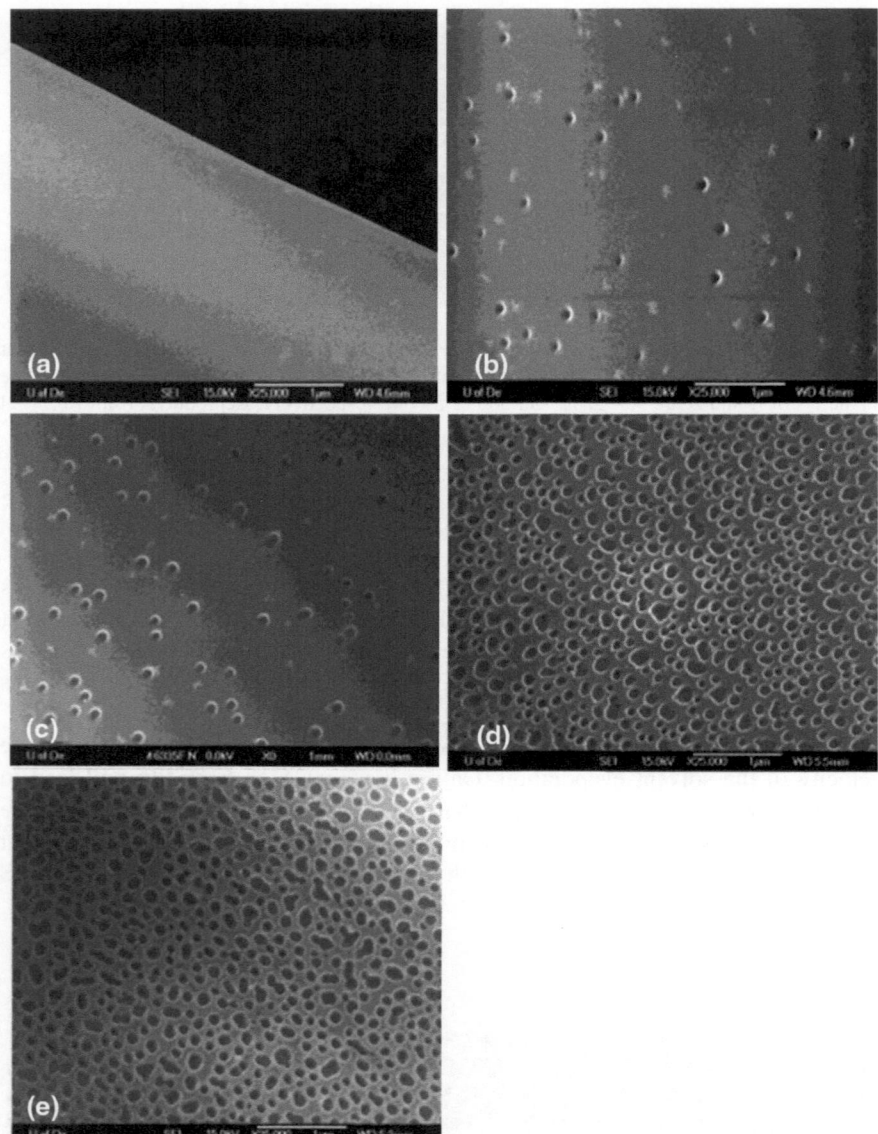

Fig. 2.16 FESEM images of surface morphologies of electrospun PS fibers under varying humidity: **a** <25 %, **b** 31–38 %, **c** 40–45 %, **d** 50–59 %, **e** 60–72 %. Reprinted with the permission from Ref. [32]. Copyright 2004 American Chemical Society

References

1. Deitzel JM, Kleinmeyer J, Harris D, Beck Tan NC (2001) The effect of processing variables on the morphology of electrospun nanofibers and textiles. Polymer 42(1):261–272. doi:10.1016/s0032-3861(00)00250-0
2. Eda G, Shivkumar S (2007) Bead-to-fiber transition in electrospun polystyrene. J Appl Polym Sci 106(1):475–487. doi:10.1002/app.25907
3. Fong H, Chun I, Reneker DH (1999) Beaded nanofibers formed during electrospinning. Polymer 40(16):4585–4592. doi:10.1016/s0032-3861(99)00068-3
4. Lee KH, Kim HY, Bang HJ, Jung YH, Lee SG (2003) The change of bead morphology formed on electrospun polystyrene fibers. Polymer 44(14):4029–4034. doi:10.1016/s0032-3861(03)00345-8
5. Yang Q, Li Z, Hong Y, Zhao Y, Qiu S, Wang C, Wei Y (2004) Influence of solvents on the formation of ultrathin uniform poly(vinyl pyrrolidone) nanofibers with electrospinning. J Polym Sci, Part B: Polym Phys 42(20):3721–3726. doi:10.1002/polb.20222
6. Koski A, Yim K, Shivkumar S (2004) Effect of molecular weight on fibrous PVA produced by electrospinning. Mater Lett 58 (3–4):493–497. doi:http://dx.doi.org/10.1016/S0167-577X(03)00532-9
7. Zhao YY, Yang QB, Lu XF, Wang C, Wei Y (2005) Study on correlation of morphology of electrospun products of polyacrylamide with ultrahigh molecular weight. J Polym Sci, Part B: Polym Phys 43(16):2190–2195. doi:10.1002/polb.20506
8. McKee MG, Layman JM, Cashion MP, Long TE (2006) Phospholipid nonwoven electrospun membranes. Science 311(5759):353–355. doi:10.1126/science.1119790
9. Larrondo L, St. John Manley R (1981) Electrostatic fiber spinning from polymer melts. I. Experimental observations on fiber formation and properties. J Polym Sci: Polym Phys Ed 19(6):909–920. doi:10.1002/pol.1981.180190601
10. Sukigara S, Gandhi M, Ayutsede J, Micklus M, Ko F (2003) Regeneration of Bombyx mori silk by electrospinning—part 1: Processing parameters and geometric properties. Polymer 44(19):5721–5727. doi:10.1016/s0032-3861(03)00532-9
11. Ding B, Kim H-Y, Lee S-C, Shao C-L, Lee D-R, Park S-J, Kwag G-B, Choi K-J (2002) Preparation and characterization of a nanoscale poly(vinyl alcohol) fiber aggregate produced by an electrospinning method. J Polym Sci, Part B: Polym Phys 40(13):1261–1268. doi:10.1002/polb.10191
12. Ki CS, Baek DH, Gang KD, Lee KH, Um IC, Park YH (2005) Characterization of gelatin nanofiber prepared from gelatin–formic acid solution. Polymer 46(14):5094–5102. doi:10.1016/j.polymer.2005.04.040
13. Kim K-H, Jeong L, Park H-N, Shin S-Y, Park W-H, Lee S-C, Kim T-I, Park Y-J, Seol Y-J, Lee Y-M, Ku Y, Rhyu I-C, Han S-B, Chung C-P (2005) Biological efficacy of silk fibroin nanofiber membranes for guided bone regeneration. J Biotechnol 120(3):327–339. doi:10.1016/j.jbiotec.2005.06.033
14. Lee JS, Choi KH, Ghim HD, Kim SS, Chun DH, Kim HY, Lyoo WS (2004) Role of molecular weight of atactic poly(vinyl alcohol) (PVA) in the structure and properties of PVA nanofabric prepared by electrospinning. J Appl Polym Sci 93(4):1638–1646. doi:10.1002/app.20602
15. Zhang Y, Ouyang H, Lim CT, Ramakrishna S, Huang Z-M (2005) Electrospinning of gelatin fibers and gelatin/PCL composite fibrous scaffolds. J Biomed Mater Res Part B: Appl Biomater 72B(1):156–165. doi:10.1002/jbm.b.30128
16. Haghi AK, Akbari M (2007) Trends in electrospinning of natural nanofibers. Phys Status Solidi (a) 204(6):1830–1834. doi:10.1002/pssa.200675301
17. Pham QP, Sharma U, Mikos AG (2006) Electrospun Poly(ε-caprolactone) Microfiber and Multilayer Nanofiber/Microfiber Scaffolds: Characterization of Scaffolds and Measurement of Cellular Infiltration. Biomacromolecules 7(10):2796–2805. doi:10.1021/bm060680j

18. Zhang C, Yuan X, Wu L, Han Y, Sheng J (2005) Study on morphology of electrospun poly(vinyl alcohol) mats. Eur Polym J 41(3):423–432. doi:10.1016/j.eurpolymj.2004.10.027
19. Zong X, Kim K, Fang D, Ran S, Hsiao BS, Chu B (2002) Structure and process relationship of electrospun bioabsorbable nanofiber membranes. Polymer 43(16):4403–4412. doi:10.1016/s0032-3861(02)00275-6
20. Huang C, Chen S, Lai C, Reneker Darrell H, Qiu H, Ye Y, Hou H (2006) Electrospun polymer nanofibres with small diameters. Nanotechnology 17(6):1558–1563. doi:10.1088/0957-4484/17/6/004
21. Reneker DH, Chun I (1996) Nanometre diameter fibres of polymer, produced by electrospinning. Nanotechnology 7(3):216–223. doi:10.1088/0957-4484/7/3/009
22. Yuan X, Zhang Y, Dong C, Sheng J (2004) Morphology of ultrafine polysulfone fibers prepared by electrospinning. Polym Int 53(11):1704–1710. doi:10.1002/pi.1538
23. Buchko CJ, Chen LC, Shen Y, Martin DC (1999) Processing and microstructural characterization of porous biocompatible protein polymer thin films. Polymer 40(26):7397–7407. doi:10.1016/s0032-3861(98)00866-0
24. Demir MM, Yilgor I, Yilgor E, Erman B (2002) Electrospinning of polyurethane fibers. Polymer 43(11):3303–3309. doi:10.1016/s0032-3861(02)00136-2
25. Yördem OS, Papila M, Menceloğlu YZ (2008) Effects of electrospinning parameters on polyacrylonitrile nanofiber diameter: An investigation by response surface methodology. Mater Des 29(1):34–44. doi:10.1016/j.matdes.2006.12.013
26. Wang X, Um IC, Fang D, Okamoto A, Hsiao BS, Chu B (2005) Formation of water-resistant hyaluronic acid nanofibers by blowing-assisted electro-spinning and non-toxic post treatments. Polymer 46(13):4853–4867. doi:10.1016/j.polymer.2005.03.058
27. Sundaray B, Subramanian V, Natarajan TS, Xiang R-Z, Chang C–C, Fann W-S (2004) Electrospinning of continuous aligned polymer fibers. Appl Phys Lett 84(7):1222–1224
28. Li D, Wang Y, Xia Y (2004) Electrospinning nanofibers as uniaxially aligned arrays and layer-by-layer stacked films. Adv Mater 16(4):361–366. doi:10.1002/adma.200306226
29. Xu CY, Inai R, Kotaki M, Ramakrishna S (2004) Aligned biodegradable nanofibrous structure: A potential scaffold for blood vessel engineering. Biomaterials 25(5):877–886. doi:10.1016/s0142-9612(03)00593-3
30. Ki CS, Kim JW, Hyun JH, Lee KH, Hattori M, Rah DK, Park YH (2007) Electrospun three-dimensional silk fibroin nanofibrous scaffold. J Appl Polym Sci 106(6):3922–3928. doi:10.1002/app.26914
31. Mit-uppatham C, Nithitanakul M, Supaphol P (2004) Ultrafine electrospun polyamide-6 fibers: Effect of solution conditions on morphology and average fiber diameter. Macromol Chem Phys 205(17):2327–2338. doi:10.1002/macp.200400225
32. Casper CL, Stephens JS, Tassi NG, Chase DB, Rabolt JF (2004) Controlling surface morphology of electrospun polystyrene fibers: Effect of humidity and molecular weight in the electrospinning process. Macromolecules 37(2):573–578. doi:10.1021/ma0351975

Chapter 3
Electrospun Fibers Properties

Abstract Generally, electrospun nanofibers obtained on the collectors often existed in the form of nonwoven mats owing to the intrinsic instability. Although such nonwoven mats have gained tremendous focus in diverse applications including texturing, fiber reinforcement, tissue engineer, filter, sensing, and enzyme immobilization, the potential applications of electrospun fibers still cannot be fully realized. To overcome this obstacle, much effort is still needed to accurately control the properties of fibers in electrospinning, such as fiber direction, fiber surface morphology, fiber structure, and three-dimensional structures formed by electrospun nanofibers. In this chapter, we present comprehensive introduction and the corresponding mechanisms in detail.

Keywords Nonwoven · Fiber direction · Fiber surface morphology · Fiber structure · 3D structures

In Sect. 3.1, we depict the general route to control the fiber direction through the modification of electrospinning setup. Then, the fiber surface morphology control, fiber structure control, and 3D structures formed by electrospun nanofibers will be stated in Sects. 3.2–3.4, respectively.

3.1 Fibers Direction: Modification of Electrospinning Setup

For many applications, it is necessary to control the direction of 1D nanostructures such as electronics [1], skeletal muscle regeneration [2], and polarized luminescence [3]. In the past decade, a number of methods have been developed to control the electrospun fiber direction through the modification of electrospinning setup (*mainly concentrated in modification of the distances and collectors*).

Z. Li and C. Wang, *One-Dimensional Nanostructures*,
SpringerBriefs in Materials, DOI: 10.1007/978-3-642-36427-3_3,
© The Author(s) 2013

3.1.1 Modification of the Distance: Near-Field
Electrospinning

As we mentioned before that during the process of electrospinning, when the
charged jets ejected from the Taylor cone, the jets will pass by in a nearly straight
line preceded by bending into a complex path [4]. Prompted by such very
important clue: *a nearly straight line during the electrospinning process*, Prof. Lin
and co-workers [5] developed a novel route to control the fiber direction via
Near-field electrospinning (NFE), in which the distance between the tip of the
syringe and the collector is very short ranging from 500 μm to 3 mm (the nearly
straight line prior to the complex path).

   During the NFE process, three very important working parameters should be
noted:

(a) The distance between the tip of the syringe and the collector is in the range of
     500 μm to 3 mm to utilize the stable liquid jets' region for fiber direction
     control.
(b) The applied voltage is reduced owing to the short distance, and the electrical
     field strength is kept in the range of 10^7 V/m as conventional electrospinning.
(c) Discrete droplets of polymer solution should be supplied in a manner analo-
     gous to that of dip pen by immersing and pulling the electrode into and out of
     the polymer solution.

   To better understand the setup, Fig. 3.1 illustrates the setup of NFE.

   Currently, many groups have engaged in the development of NFE technique,
even patterned fiber mats have been obtained [6–13]. For example, Chang et al. [8]
used NFE for large area deposition of patterned nanofiber mats with PEO as model
(Fig. 3.2).

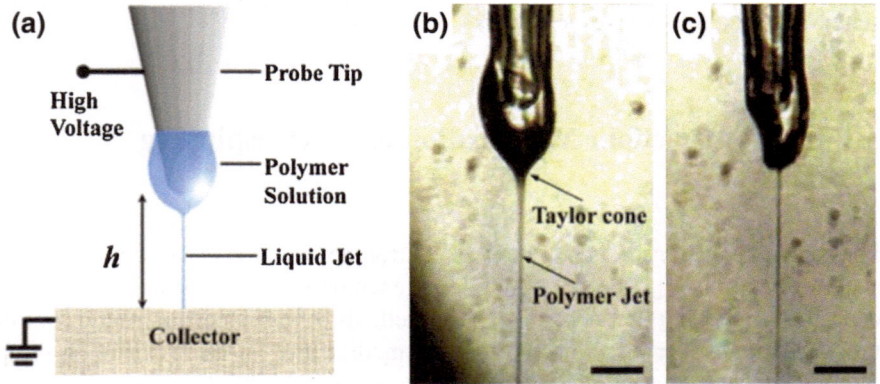

Fig. 3.1 a Schematic representation of near-field electrospinning. **b** Optical micrograph of a
polymer jet ejected from the apex of Taylor cone under applied electrical field. **c** Optical
micrograph of the polymer droplet decreases as the polymer continues to electrospinning.
Reprinted with permission from Ref. [5]. Copyright 2006 by American Physical Society

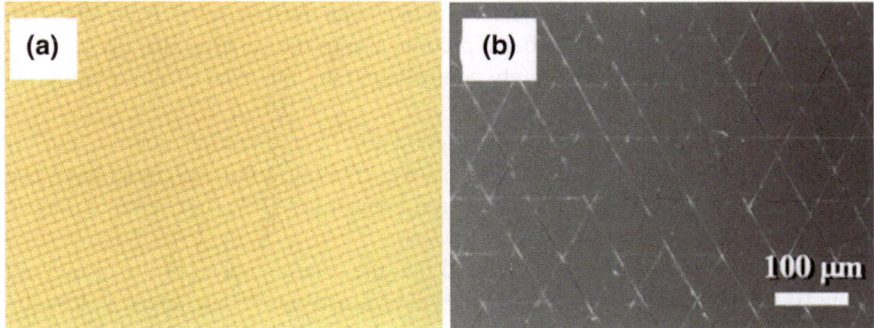

Fig. 3.2 Patterned mats obtained by NFE, in which the fiber direction can be accurately controlled. Reprinted with permission from Ref. [8]. Copyright 2008 by American Physical Society

3.1.2 Modification of Collectors

In contrast to the modification of distances, the modification of collectors for fiber direction control has been widely investigated. Till now, different types of the collectors for the fiber direction control have been successfully developed including rotating conductive collectors, counter-conductive collectors, liquid-bath collector.

3.1.2.1 Rotating Conductive Collectors

In 2001, Theron and Yarin used electrostatic field-assisted assembly technique in electrospinning for the fiber direction control [14]. In their experiment, a rotating disk has been chosen as the collector instead of the traditional collector (Al foil) for highly aligned fibers with PEO as model (Fig. 3.3a). In this chapter, the authors suggested that the aligned fiber direction was caused by the residual charges of

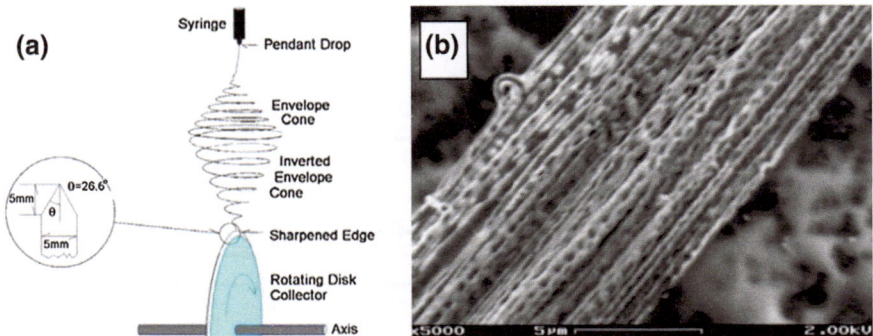

Fig. 3.3 a Schematic diagram of experimental setup with a rotating disk as collector. **b** SEM image of a braid of highly aligned PE nanofibers. Reprinted with permission from Ref. [14]. Copyright 2001 by IOP Publishing Ltd

nanofiber deposited on the collectors, which can further repel each other and thereby influence the aligned fiber direction. Using the rotating disk as collector, the fiber direction can be accurately controlled along the axial of the rotating disk (Fig. 3.3b).

Although rotating disk collector is good for fiber direction control, the collection area is really small. From 2002, novel rotating drum collectors have been developed for the large area fiber direction control [15–17]. A schematic diagram of the setup is shown in Fig. 3.4.

In this system, researchers [18] found that the rotating speed of the drum could strongly affect the degree of electrospun fiber direction. As the speed is less than 500 rpm, random mix of the electrospun fiber was collected. As the rotating speed is 4,500 rpm (approximately 1.4 m/s at the surface of the drum), the fiber direction showed significant alignment along the axial of the rotation (Fig. 3.5). If the rotating speed is too high, broken fibers will be obtained.

In 2004, Kim and his groups further investigated the effect of the rotating speed on the crystallinity of electrospun poly (ethylene terephthalate) (PET) [19]. They found that in contrast to the raw materials, the electrospun PET is amorphous (*It's very important to emphasize that such retarded crystallization was universal in electrospinning owing to the fast solvent evaporation and the rapid solidification of the stretched polymer chains at high elongational rates*). However, the peak (010) becomes more pronounced as the linear speak of the drum surface is increased, indicating that the crystals have a significant orientation caused by the linear speed (Fig. 3.6).

In the rotating conductive collector systems, one working parameter should be emphasized: The rotating speed is very important for the fiber direction control. As the rotating speed is very low, disordered nanofiber mats will be obtained. While the rotating speed is too high, fiber will break during the collection.

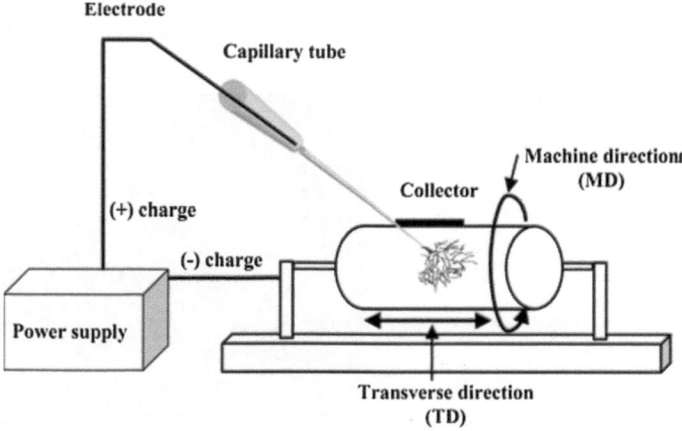

Fig. 3.4 Schematic diagram of experimental setup (*left*) and side view (*right*) of fiber direction control by using the rotating drum instead of Al foil. Reprinted from Ref. [17], with kind permission from Springer Science+Business Media

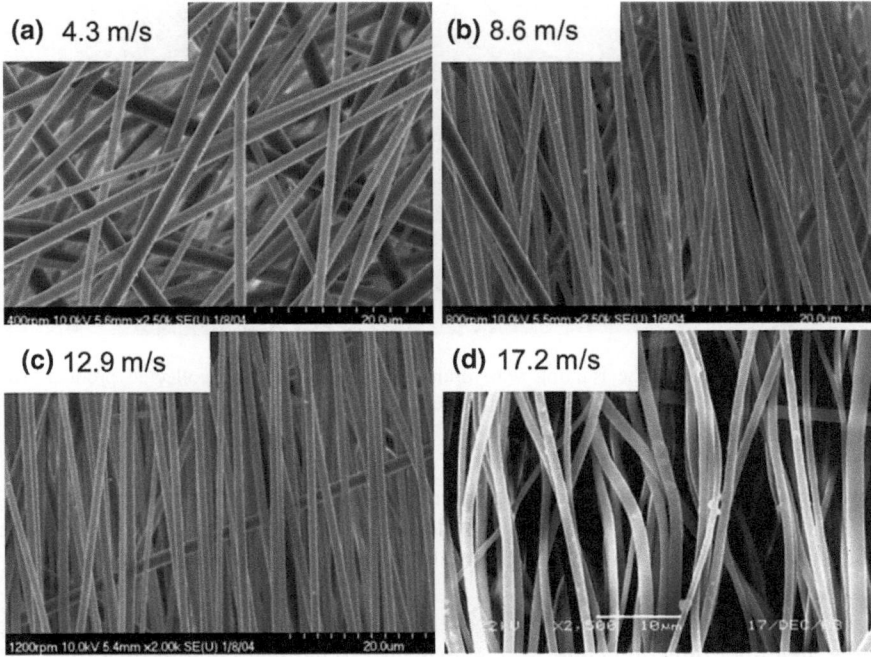

Fig. 3.5 FESEM images of the effect of the rotating speed on fiber direction control. Reproduced from Ref. [18] by permission from John Wiley & Sons Ltd

Fig. 3.6 Wide-angle X-ray diffraction (*WAXD*) of the electrospun PET nonwovens as a function of the linear speed of drum surface. Reprinted from Ref. [19], with kind permission from Springer Science+Business Media

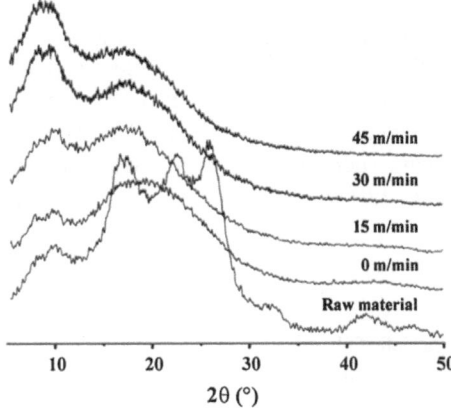

Inspired by the rotating drum collectors, Chase and Katta invented the rotating wire drum collector in 2004 [20]. Two circular nonconductive plexiglass disks 12.7 cm in diameter with a 1.2 cm hole cut in the center. Each disk has 6 mm-deep saw-cut notches placed one centimeter apart around the circumference. The two

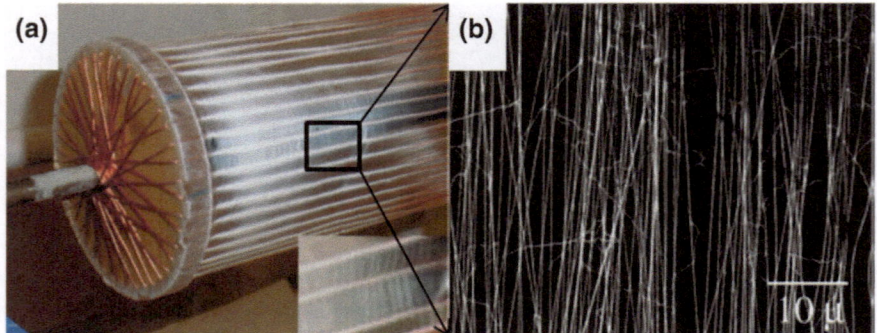

Fig. 3.7 Photograph of the rotating wire drum collector (**a**) and its collected products (**b**). Reprinted with the permission from Ref. [20]. Copyright 2004 by American Chemical Society

disks are mounted on rod and space a certain distance with PVC pipe. The conductive wire is stretched between the slots cut into the edges of the disks. The photograph of the rotating wire drum collector and its collected products has been illustrated in Fig. 3.7.

Despite those successes obtained by the rotating conductor collectors for the large area fiber direction control, scientists also found that random fiber mats can be also obtained by increasing the collecting time owing to the residual charge accumulation on the deposited fibers, which can interfere with the incoming fibers [14]. To overcome this problem, a possible way is to reduce the charges on the rotating conductive collectors. For this intent, Pan et al. [21] developed a novel electrospinning setup containing two counter-electrodes and rotating conductive collector (Fig. 3.8). In this electrospinning setup, two electrodes (syringe) were installed in opposite direction. Positive and negative voltages were applied to the two electrodes, respectively. A rotating conductive drum was used as the collector. Using such electrospinning setup, the fiber yarn direction has greatly improved (Fig. 3.9).

Fig. 3.8 Schematic electrospinning setup containing two counter-electrodes and rotating conductive collector. Reprinted with permission from Ref. [21]. Copyright 2006 by Elsevier

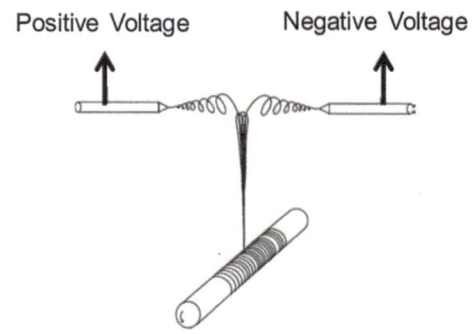

Positive Voltage   Negative Voltage

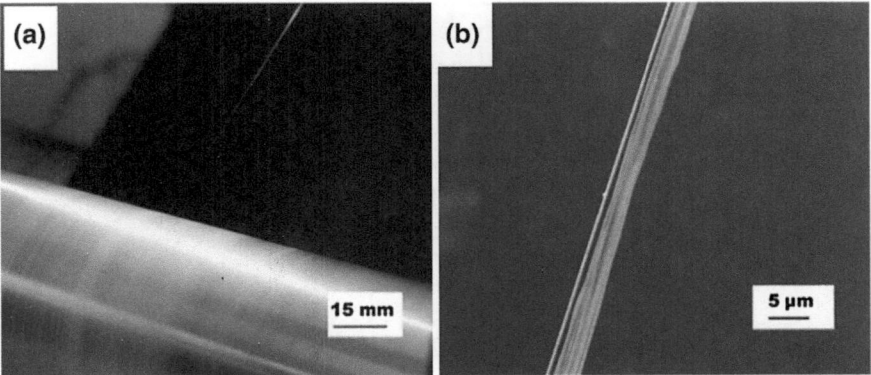

Fig. 3.9 Photograph of the fiber yarn (**a**) and SEM image of the fiber yarn (**b**). The polymer is PVA, the concentration is 13 wt. %, and the surface speed is 4.3 m/s. Reprinted with permission from Ref. [21]. Copyright 2006 by Elsevier

3.1.2.2 Counter-Conductive Collectors

Although the challenge of fiber direction control has been met to a certain degree by using the rotating conductive collectors, the electrospinning setups are really too complicated. From 2003, Li et al. [22, 23] developed a series of simple and novel electrospinning setups for the fiber direction control through counter-conductive collectors. Comparing with the traditional collectors (Al foils), a void gap has been introduced between the two collectors (Fig. 3.10). Figure 3.10a illustrates the schematic setup for the counter-conductive collectors. Figure 3.10b shows a cross-sectional view of the electric field strength vectors between the syringe and the grounded collectors. Figure 3.10c displays the electrostatic forces acting on a charged considered to be a string of positively charged fiber connected through the viscoelastic medium, in which the charged nanofibers experience two sets of electrostatic forces: the first set (F_1) originating from the splitting electric field and second one (F_2) between the charged fiber and image charges induced on the surfaces of the two grounded electrodes. Additionally, unlike fibers directly deposited on the top of an electrode where they can be fast neutralized, the fiber suspended across the gap can retain large charges. Thus, the electrostatic repulsion between the deposited and the upcoming fibers can further enhance the parallel alignment.

In the counter-conductive collector systems, several important working parameters should be emphasized.

1. When the distance between the gaps is longer than 1.5 cm, nanofibers, with the diameter thinner than 150 nm, tend to be broken during the electrospinning for those fibers were not sufficient strong to support their own weight, as well as the electrostatic repulsion applied by other charged fibers.
2. As the distance is fixed (>2 cm), the conductivity of the collectors is very important in determining the fiber direction control. Generally, fiber direction

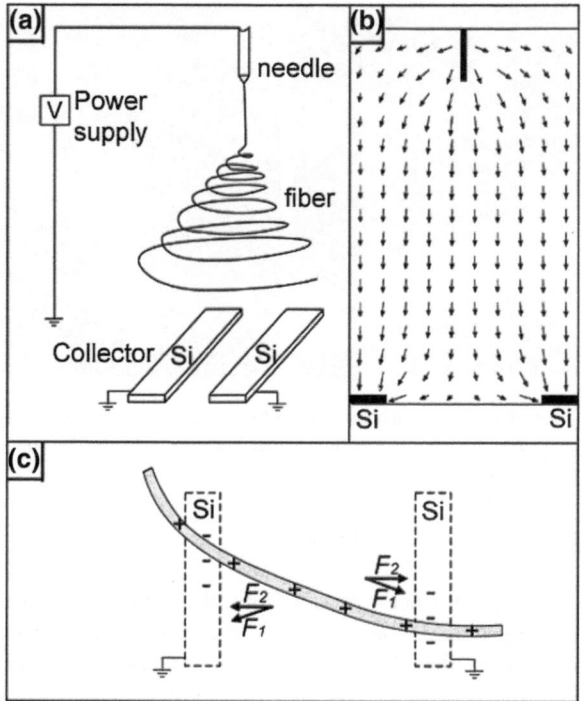

Fig. 3.10 **a** Schematic diagram of the setup of the counter-conductive collectors. **b** Calculated electric field strength vectors between the syringe and counter-conductive collectors. **c** Electrostatic force between the gaps. Reprinted with the permission from Ref. [22]. Copyright 2003 by American Chemical Society

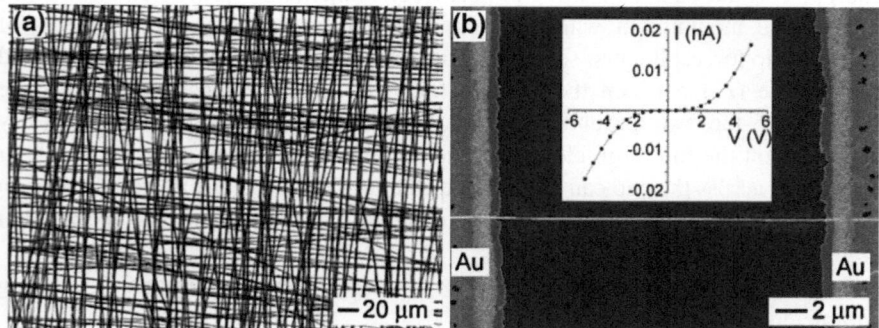

Fig. 3.11 **a** Optical micrograph of grid formed by stacking two layers of aligned nanofibers. **b** SEM image of simple device structures fabricated in this method. The inset in **b** shows a typical I–V curve measured from a single Sb-doped SnO_2 nanofiber. Reprinted with the permission from Ref. [22]. Copyright 2003 by American Chemical Society

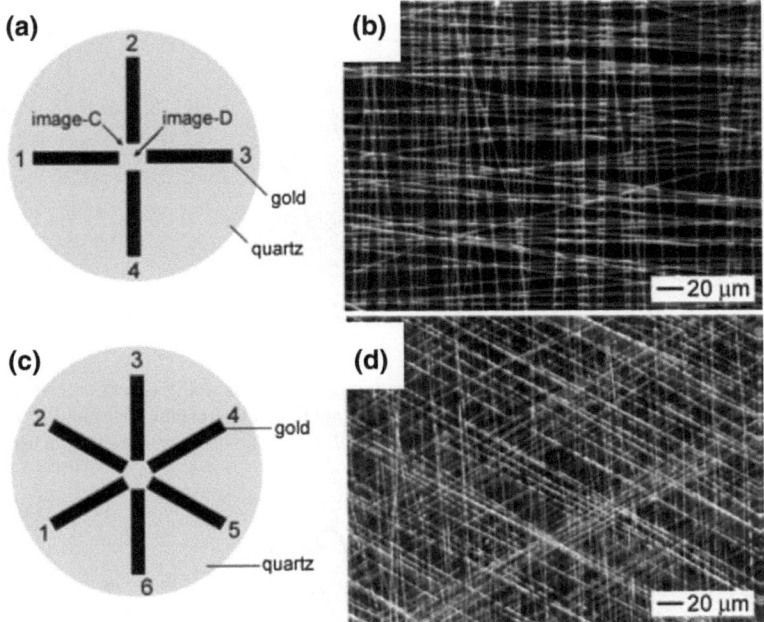

Fig. 3.12 Schematic illustration of the electrospinning setup containing four (**a**) and six (**c**) counter-conductive collectors (Au electrodes deposited on quartz wafers). **b** and **d** are optical micrographs of the fibers collected in the center area of the gold electrodes. Reproduced from Ref. [23] by permission from John Wiley & Sons Ltd

of the electrospun fibers collected with conductive collectors is better than that of the electrospun fibers collected with insulated collectors (e.g., glass slides, polymer plates, and silicon wafer).

Unique advantages in the counter-conductive collector systems:

(1) Complex fiber direction control (3D grid structure) can be realized by stacking in a layer-by-layer fashion (Fig. 3.11a).
(2) In addition to arrays containing large numbers of fibers, well-separated nanofibers could be also obtained by shortening the collection time (Fig. 3.11b), which makes the fibers particularly useful as building blocks in fabricating device structures.

In 2004, four counter-conductive collectors and six counter-conductive collectors have been developed for the fiber complex direction control (Fig. 3.12) [23].

Prompted by the Li and Xia's work, Teo and Ramakrishna [24] used two counter-conductive blades as the collectors (Fig. 3.13a). Comparing with the previous systems, the area of the fiber depositions has been reduced and fixed; thus, highly aligned fibers can be obtained between the two counter-conductive blades (Fig. 3.13b).

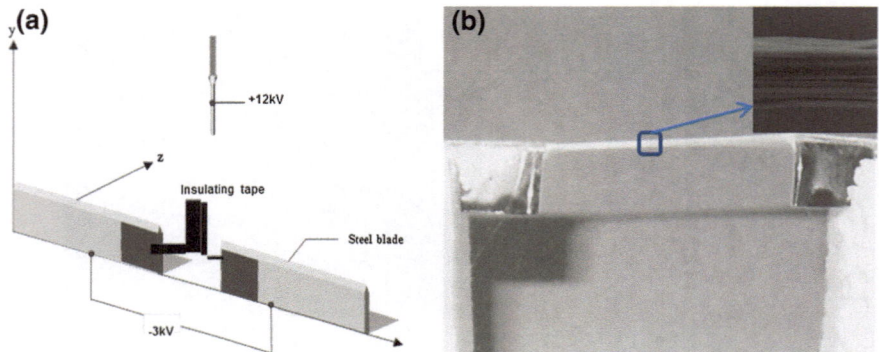

Fig. 3.13 a Schematic diagram of the electrospinning setup containing two counter-conductive blades as the collectors. **b** Optical micrograph of the obtained electrospun fiber bundle. The inset in **b** is the SEM image of the electrospun fibers. Reprinted with permission from Ref. [24]. Copyright 2005 by IOP Publishing Ltd

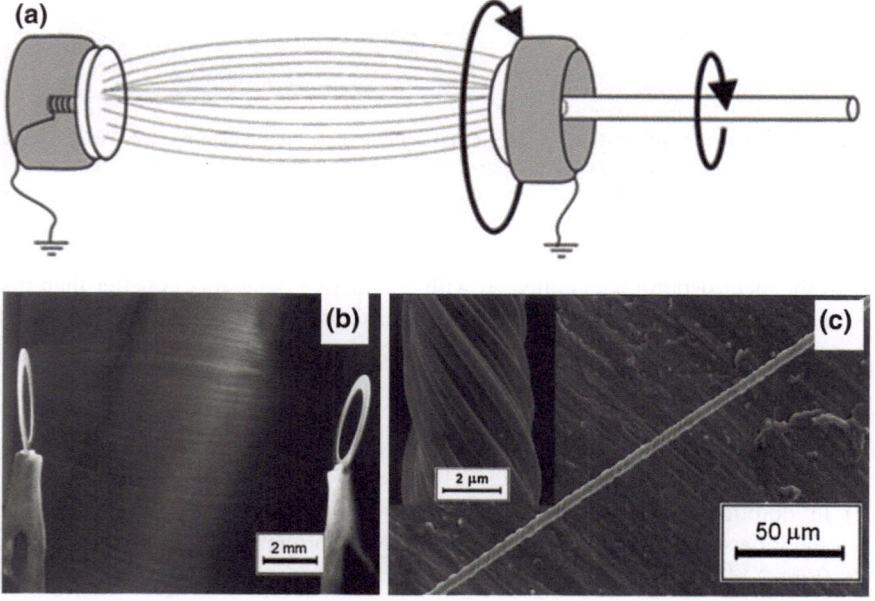

Fig. 3.14 a Schematic diagram of the electrospinning setup containing two parallel rings. **b** Optical micrographs of the electrospun fibers across the parallel rings. **c** SEM images of the electrospun fiber yarn. Reprinted with permission from Ref. [25]. Copyright 2005 by Elsevier

In 2005, Dalton and Möller established another type of counter-conductive collector. They used two parallel rings as the counter collectors (Fig. 3.14a) [25]. Those two parallel rings are good for the fiber yarn direction control (Fig. 3.14b and c).

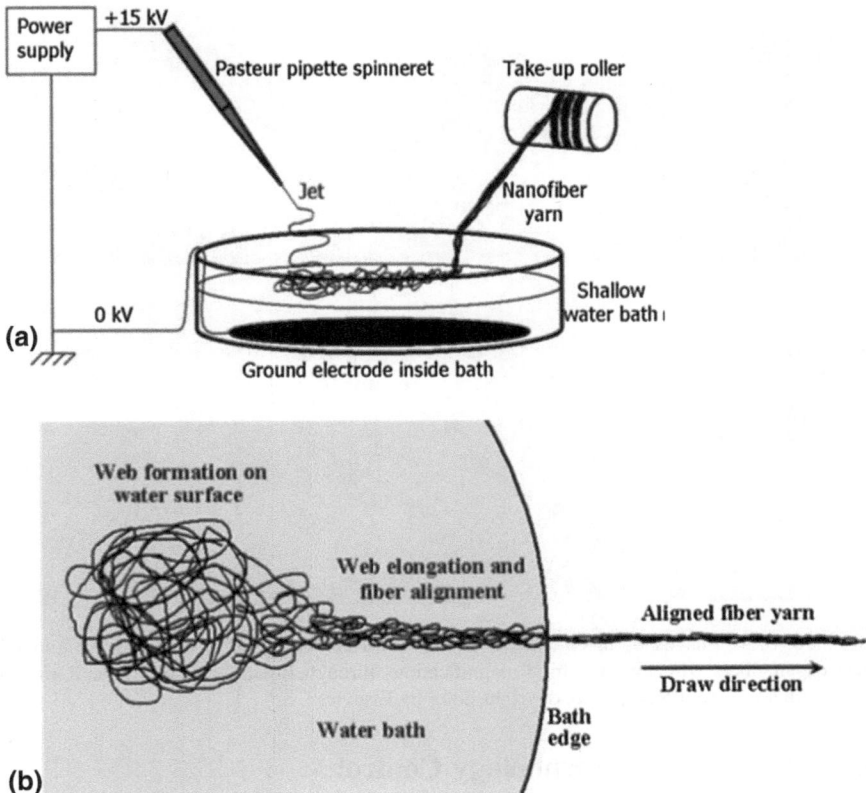

Fig. 3.15 **a** Schematic diagram of the electrospinning setup containing liquid-bath collectors. **b** Schematic diagram of the aligned fiber formation process. Reprinted with permission from Ref. [27]. Copyright 2005 by Elsevier

3.1.2.3 Liquid-Bath Collectors

In 2005, Smit and Khil simultaneously reported a novel method for fiber direction control by using liquid bath (water reservoir) as the collector (Fig. 3.15a) [26, 27]. The mechanism can be explained as follows: (1) A flat web of fibers forms on the surface of the water bath with randomly looped fibers. (2) During the drawing process for transferring the fiber, the web is elongated, and alignment for the fibers takes place. (3) The surface tension of the remaining water on the web pulls the fibers together into a 3D round yarn structure (Fig. 3.15b). Using this method, the fiber direction control of different type of polymers can be realized in some degree (Fig. 3.16).

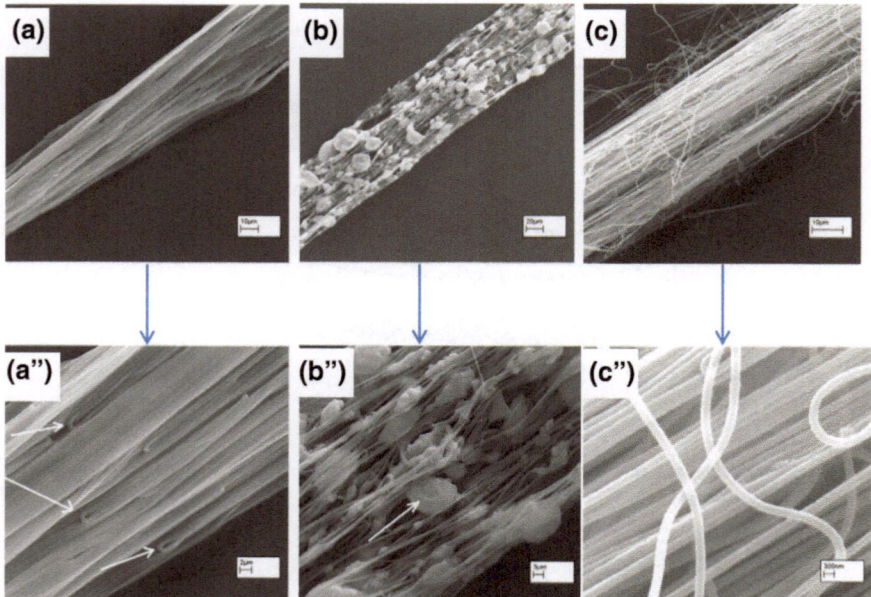

Fig. 3.16 SEM images of the PVAc (**a** and **a"**), PVDF (**b** and **b"**), and PAN (**c** and **c"**) electrospun fibers with low and high magnifications through liquid-bath collectors. Reprinted with permission from Ref. [27]. Copyright 2005 by Elsevier

3.2 Fiber Surface Morphology Control

In general, polymer nanofiber obtained by electrospinning technique exhibits solidified nanofibers with smooth surface. For selected applications, it is desirable to control not only the fiber diameter, but also fiber surface morphology. Generally, fiber surface morphology control was mainly focused on constructing pores and lines on surface.

3.2.1 Pores on Surfaces

In 2001, Bognitzki et al. [28] successfully build pores on the surface of electrospun polymer (PLLA, PC, and polyvinylcarbazole) nanofibers with dichloromethane (Fig. 3.17). In their study, dichloromethane with the very low boiling point of 40 °C was chosen as solvent. The reason for pores on the surface of electrospun fiber is induced by the fast evaporation of the solvent and subsequent rapid solidification.

From 2002, Rabolt [29, 30] and co-workers successfully build pores on the surface of electrospun PS fibers with the aid of solvent volatility (Fig. 3.18a) or humidity (Fig. 3.18b). They also investigated the effect of humidity on the pore

Fig. 3.17 SEM images of the different type of polymer electrospun fiber with pores on surface. Reproduced from Ref. [28] by permission from John Wiley & Sons Ltd

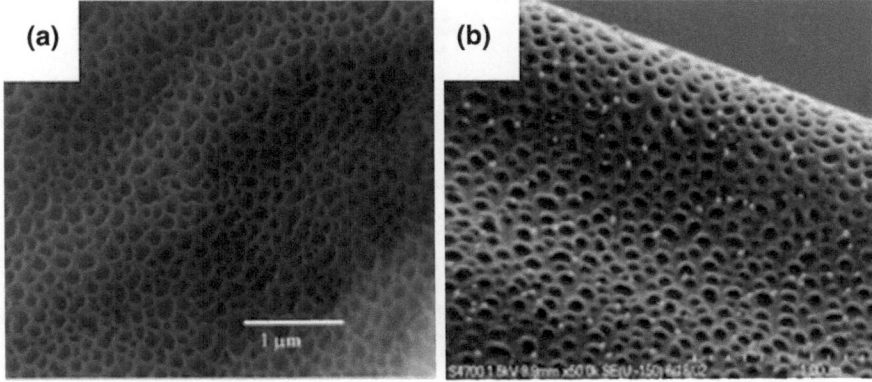

Fig. 3.18 FESEM micrographs of PS/THF fiber from tetrahydrofuran (*THF*) solvent (**a**) and electrospun in 50 % humidity (**b**), respectively. Reprinted with the permission from Ref. [29]. Copyright 2002 by American Chemical Society. Reprinted with the permission from Ref. [30]. Copyright 2004 by American Chemical Society

size. They found that increasing the humidity, the pore size becomes larger (Fig. 3.19). They suggested that the formation of pore is phase separation and four main phase separations are as follows: thermally induced phase separation, immersion precipitation, air-casting of the polymer solution, and precipitation from the vapor phase [31, 32].

3.2.2 Parallel Line on Surfaces

Recently, Lin and Huang [33, 34] construct parallel line on the surface of electrospun cellulose acetate butyrate (CAB) fiber from the acetone/DMAc solution. They investigated the effect of the solution concentration of the formation on parallel line on the surface (Fig. 3.20). They found that as the CAB concentration is less than 12 wt. %, no parallel line can be obtained.

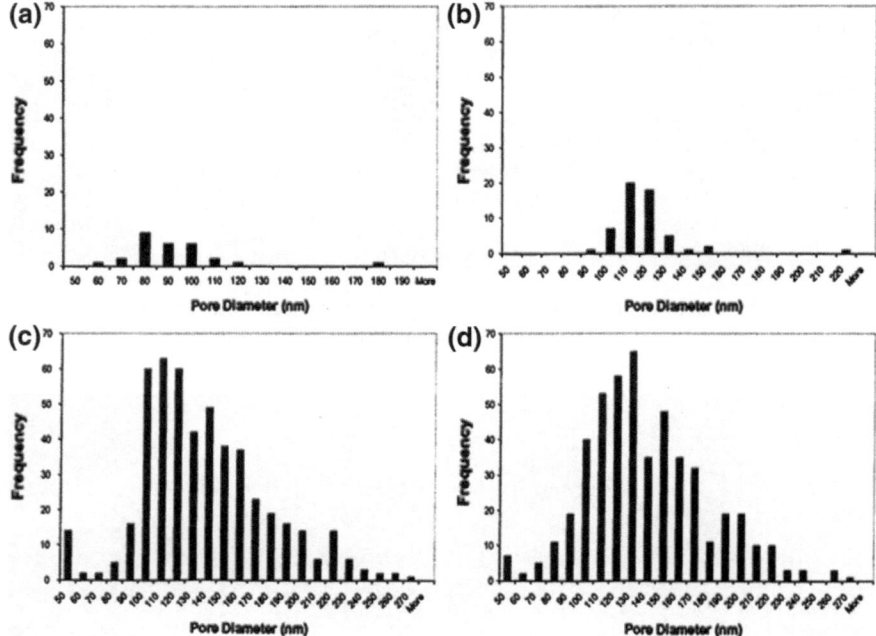

Fig. 3.19 Pore diameter distributions of pores based on PS (MW: 190,000 g/mol) electrospun fiber at varying humidity ranges: **a** 31–38 %, **b** 40–45 %, **c** 50–59 %, **d** 60–72 %. Reprinted with the permission from Ref. [30]. Copyright 2004 by American Chemical Society

They suggested that the formation of parallel line on the surface of CAB fiber was caused by the phase separation (Fig. 3.21). (a) Voids are initially formed on the jet surface at the early electrospinning stage due to the strong convection and instantaneous evaporation of acetone from the filament. (b) With elongation of the filaments, the voids are highly stretched to form oriented lines on the filament surface. (c) The complete evaporation of the solvent at the final stage of electrospinning results in solid fibers with the line surface texture fixed.

3.3 Fiber Structure Control

3.3.1 Fiber Structure Control by Co-Electrospinning

3.3.1.1 Core–Shell Structures

In electrospinning field, a notable breakthrough is the invention of co-electrospinning technique in which two or multi-coaxial capillaries have been used instead of the traditional single spinneret; thus, two or multi-fluids can be used for core–shell

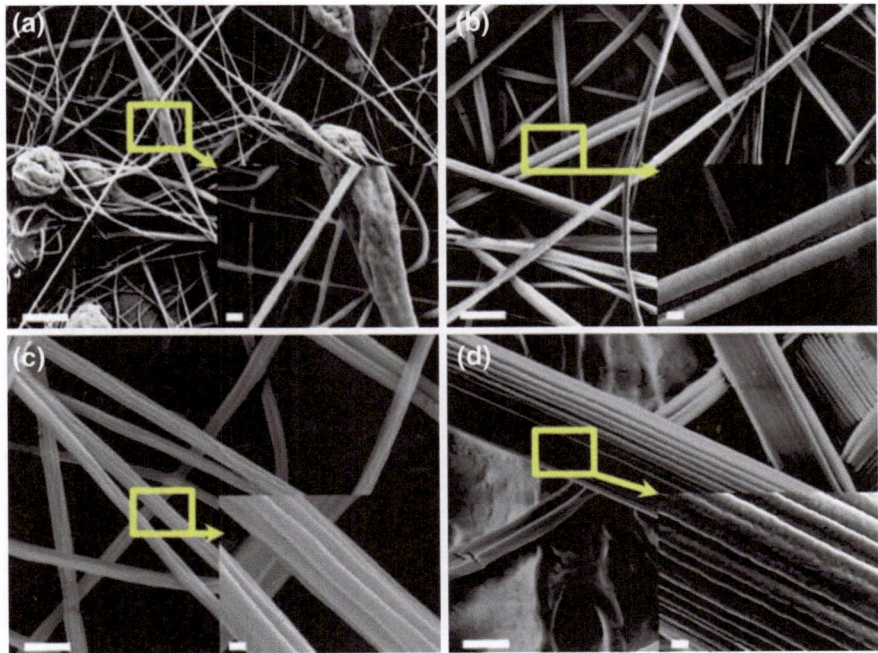

Fig. 3.20 SEM images of the CAB fiber from different concentrations: **a** 5 wt. %, **b** 12 wt. %, **c** 15 wt. %, and **d** 30 wt. %. Solvent: acetone/DMAc = 2/1, v.v. Reproduced from Ref. [34] by permission from The Royal Society of Chemistry (*RSC*) on behalf of the Centre National de la Recherche Scientifique (*CNRS*) and the RSC

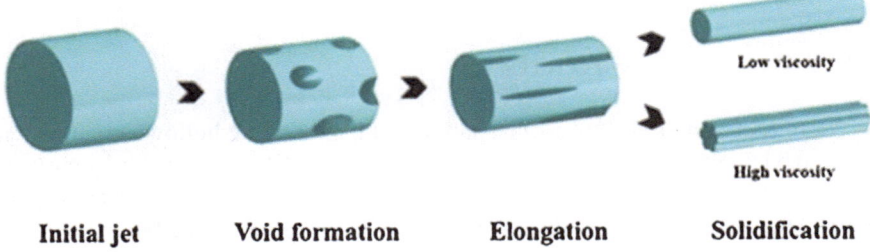

Low viscosity

High viscosity

Initial jet   **Void formation**   **Elongation**   **Solidification**

Fig. 3.21 Schematic diagram of the parallel line formation on the surface. Reproduced from Ref. [34] by permission from The Royal Society of Chemistry (*RSC*) on behalf of the Centre National de la Recherche Scientifique (*CNRS*) and the RSC

or more complicated compound jet in electric field and then solidified to desirable structures.

In 2003, Sun et al. [35] firstly developed a novel electrospinning setup entitled "co-electrospinning," in which the traditional single syringe was taken place by two coaxial capillaries (Fig. 3.22). Till now, diverse compound core–shell-structured fibers have been successfully obtained (Fig. 3.23) [36–44].

Fig. 3.22 Schematic illustration of the co-electrospinning setup. The spinneret was fabricated from two coaxial capillaries, through different materials that were simultaneously ejected to form a continuous, coaxial jet. Reproduced from Ref. [35] by permission from John Wiley & Sons Ltd

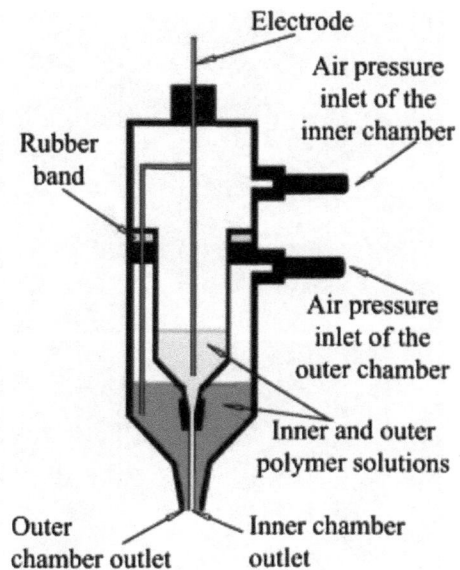

3.3.1.2 Hollow Structures

In 2004, Li and Xia [45] extended the usage of the co-electrospinning and successfully obtained hollow ceramic nanofibers (Fig. 3.24), in which mineral oil was used as the core material. After extracting and calcination, hollow ceramic nanofibers can be obtained. This technique immensely expands the variety of electrospun structures.

3.3.1.3 Bamboo-Like Structures

In 2006, McCann et al. [46] further fabricated bamboo-like hollow structures via melt coaxial electrospinning (Fig. 3.25a) with TiO_2-PVP as model. In contrast to the normal co-electrospinning setup, the inner solution is not continuous but discrete. After the extraction of the inner solution, large space will be left, resulting in the formation of bamboo-like structures (Fig. 3.25b).

3.3.1.4 Multi-Channel Structures

In 2007, Zhao et al. [47] further modified the co-electrospinning setup (Fig. 3.26a) and synthesized multi-channel microtubes within fibers (Fig. 3.26b). In this study, controllable multi-capillaries are not assembled coaxial within the spinneret. Two immiscible viscous liquids were fed separately to the three inner capillaries and the outer syringe in a suitable flow rate.

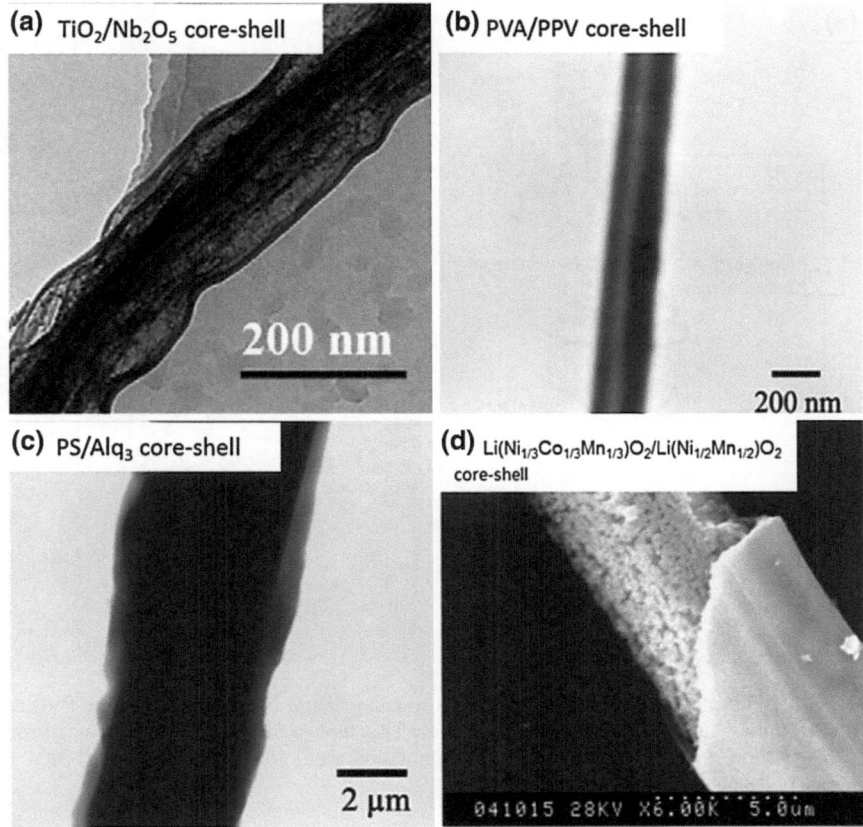

Fig. 3.23 Diverse compound core–shell-structured fibers obtained by co-electrospinning. **a** Reprinted with permission from Ref. [42]. Copyright 2012 by Elsevier. **b** and **c** Reprinted with permission from Ref. [43]. Copyright 2008 by Elsevier. **d** Reprinted with permission from Ref. [44]. Copyright 2010 by Elsevier

3.3.1.5 Fiber-in-Tube Structures

In 2010, Chen et al. [48] fabricated a novel electrospun structures: nanowire-in-microtube-structured core/shell fibers via multi-fluidic co-electrospinning (Fig. 3.27a). In this work, three coaxial capillaries were assembled as the spinneret, and a chemically inert middle fluid was introduced to work as a spacer between the outer and inner fluids. Thus, three-layered complex structures can be obtained (Fig. 3.27b).

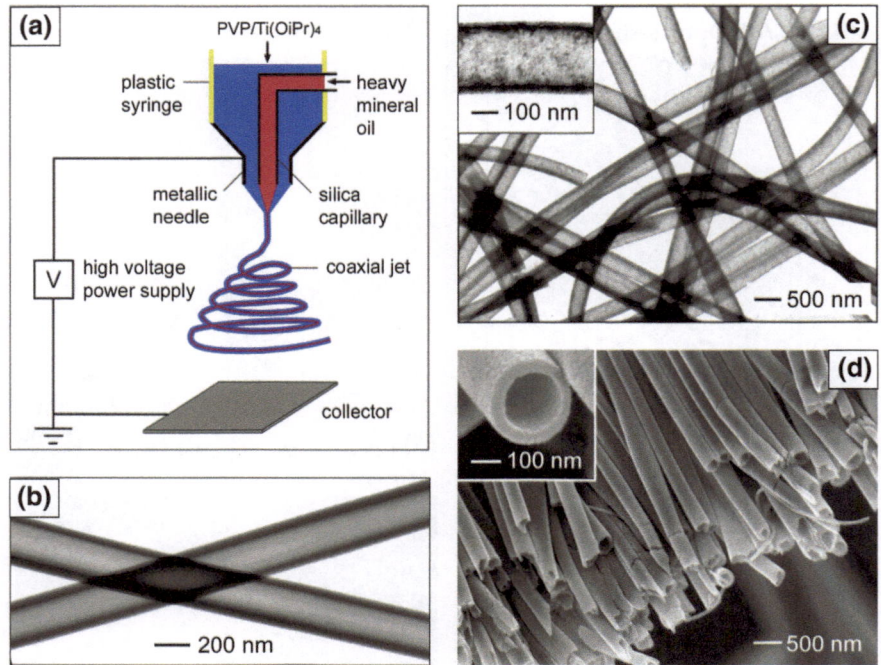

Fig. 3.24 **a** Schematic illustration of co-electrospinning setup. **b** TEM image of the PVP and Ti(OiPr)$_4$ hollow tube. **c** TEM image of the hollow TiO$_2$ tube. **d** SEM image of the aligned array of TiO$_2$ hollow tubes. Reprinted with the permission from Ref. [45]. Copyright 2004 by American Chemical Society

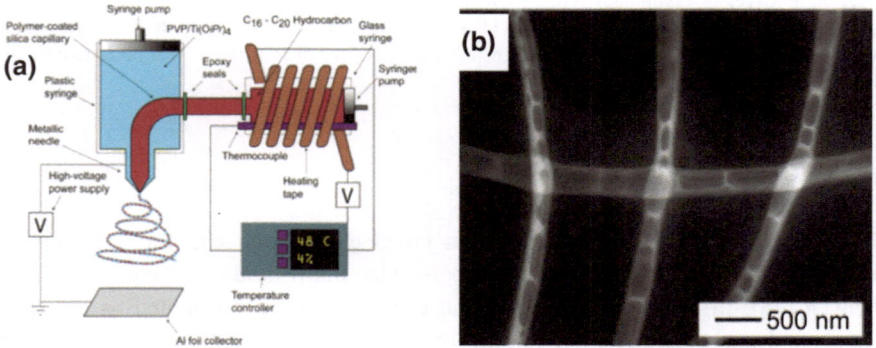

Fig. 3.25 **a** Schematic illustration of the metal co-electrospinning. **b** TEM image of the TiO$_2$-PVP bamboo-like structures. Reprinted with the permission from Ref. [46]. Copyright 2006 by American Chemical Society

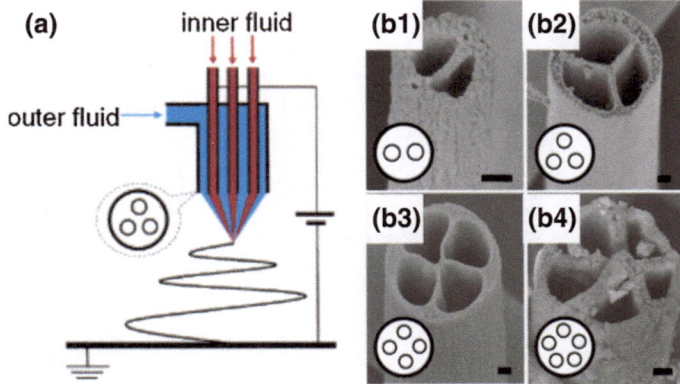

Fig. 3.26 a Schematic illustration of the modified co-electrospinning setup. **b** Controllable multi-channel microtubes within fibers. Reprinted with the permission from Ref. [47]. Copyright 2007 by American Chemical Society

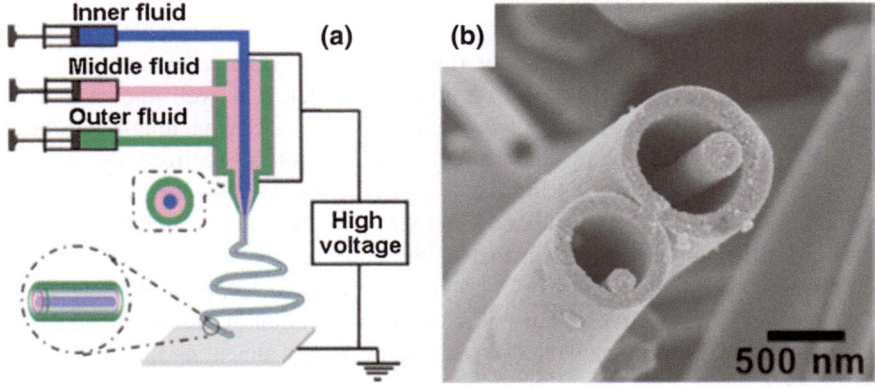

Fig. 3.27 a Schematic diagram of the multi-fluidic co-electrospinning setup. **b** Cross-sectional SEM image of the final products. Reprinted with the permission from Ref. [48]. Copyright 2010 by American Chemical Society

3.3.1.6 Side-by-Side Structures

Additionally, it should be also worth noting that Liu et al. [49] invented side-by-side dual spinneret method (Fig. 3.28a). This method is useful for the fabrication of side-by-side bicomponent nanofibers (Fig. 3.28b).

3.3.1.7 Internal Complex Structures

In 2006, interior complex structures within the electrospun fibers have realized, for the first time, by Rutledge and groups [50] with poly (styrene–isoprene–styrene)

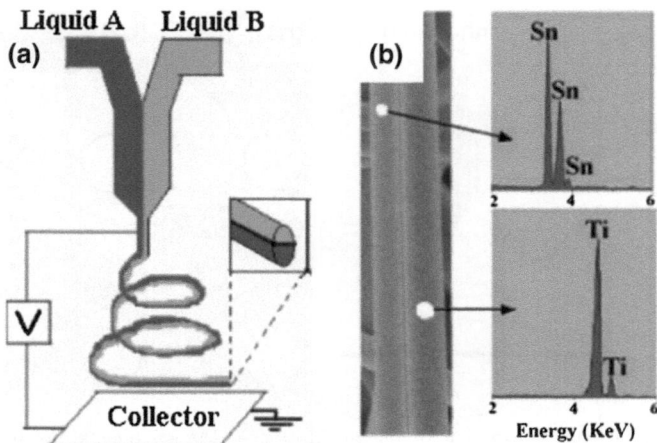

Fig. 3.28 a Schematic illustration of side-by-side dual spinneret. **b** SEM image (*left*) and EDS analysis (*right*) of the side-by-side TiO$_2$/SnO$_2$ bicomponent fibers. Reprinted with the permission from Ref. [49]. Copyright 2007 by American Chemical Society

(PSIS) block copolymer (core)/poly (methyl methacrylate-*co*-methacrylic acid) P(MMA-ran-MMA) (shell)-structured fibers as model. It is well known that the electrospun fibers with well-formed internal structures having long-range order through phase separation of block copolymer cannot be realized for the fast evaporation of the solvent during the electrospinning process. To overcome this problem, Rutledge and groups choose P (MMA-ran-MMA) as shell for it higher glass transition temperature than that of the core materials. The glass transition temperature of P (MMA-ran-MMA), styrene block, and isoprene block is 167.5, 105, and 70 °C, respectively. Thus, microphase separation of the block copolymer under cylindrical confinement can be realized by simply annealing the fibers above the upper T_g of the core materials but below the T_g of the shell polymer (Fig. 3.29).

3.3.2 Fiber Structure Control by Single Syringe

3.3.2.1 Core–Shell Structures Via Electrostatic Force Induction

In 2006, Li et al. [51] successfully obtained Cu/PVA core–shell-structured fibers through single-spinneret electrospinning setup (Fig. 3.30a) from the PVA-protected Cu nanoparticles aqueous solution. By tracking the behaviors of the PVA-protected Cu nanoparticles within the single spinneret under the applied voltage, they found that PVA-protected Cu nanoparticles can form compact PVA-protected Cu nanochains (Fig. 3.30b and c). Based on those observations, a plausible mechanism on the formation of the copper/PVA nanocables based on the mobility, the adjacent distances, and the electrostatic absorption (caused by the

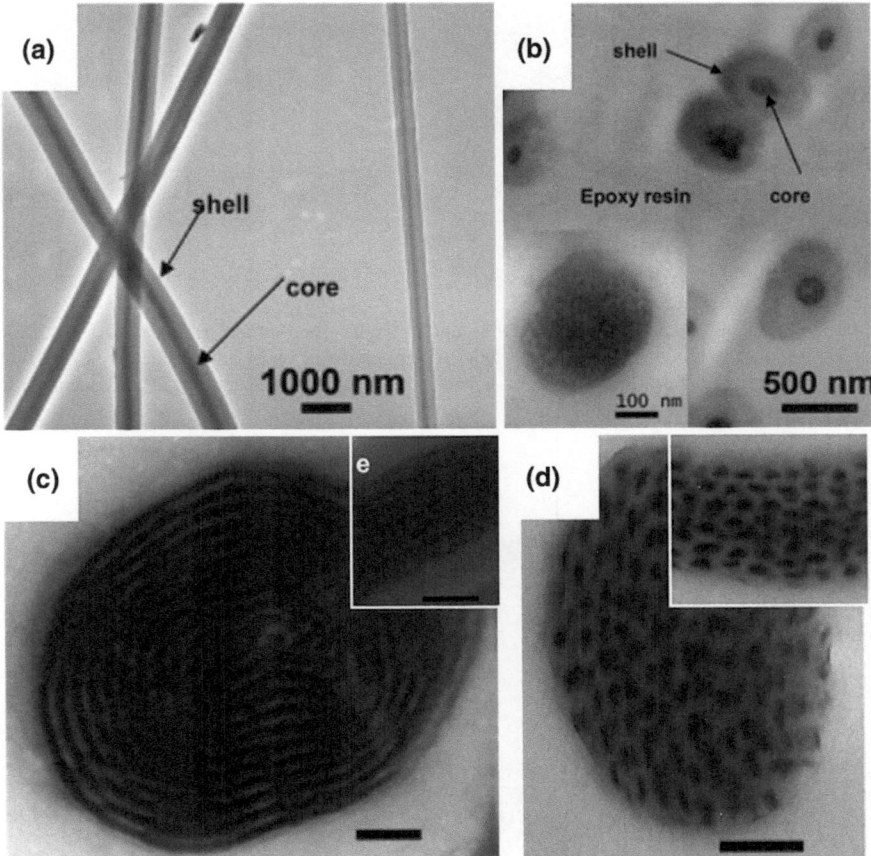

Fig. 3.29 TEM images of the core–shell structures. **a** and **b** are longitudinal view and axial view of the core–shell structures. **c** and **d** are axial view of the interior complex structures within core. Insets in **c** and **d** are corresponding longitudinal view. Reprinted with the permission from Ref. [50]. Copyright 2006 American Chemical Society

high-voltage polarization) of the PVA-protected Cu nanoparticles (PPCNs) during the electrospinning had been advanced (Fig. 3.31): 1) As the solution is polarized, the PPCNs become positively charged on one side and negatively charged on opposite side, causing the static absorption among the PPCNs (Fig. 3.31a and b). For the shorter distance between the adjacent PPCNs, compact PVA-protected Cu nanochains will form (Fig. 3.31c). Finally, Cu nanochains within the PVA nanofibers can be obtained.

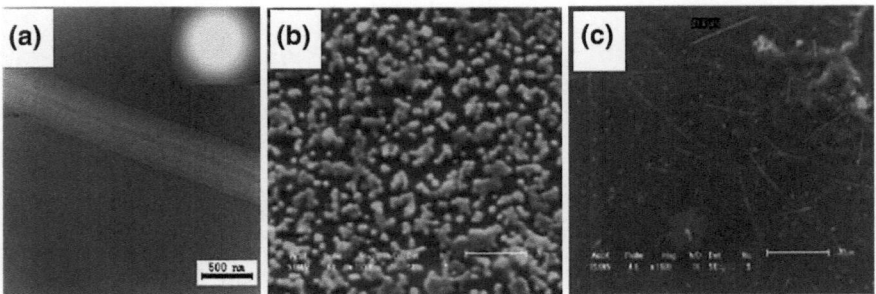

Fig. 3.30 **a** TEM image of the as-prepared Cu/PVA core–shell-structured fibers. **b** and **c** SEM images of the behaviors of PVA-protected Cu nanoparticles under the applied voltages with different polarization time. The time of **b** and **c** is 0 and 4 h, respectively. Reproduced from Ref. [51] by permission from John Wiley & Sons Ltd

Fig. 3.31 Schematic illustration on the formation of Cu/PVA core–shell-structured nanofiber. Reproduced from Ref. [51] by permission from John Wiley & Sons Ltd

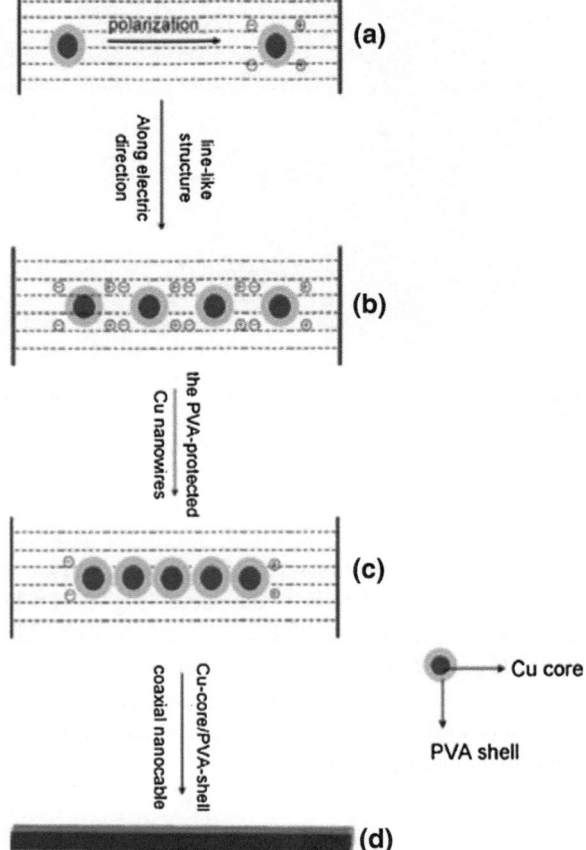

3.3.2.2 Core–Shell Structure via Emulsion Electrospinning

In 2006, Xu et al. [52] invented single-nozzle emulsion electrospinning technique for the fabrication of PEO/PEG–PLA core–shell-structured fibers (Fig. 3.32) from a water-in-oil emulsion. They suggested that the core–shell structure is formed during the electrospinning of a W/O emulsion (Fig. 3.33): (1) The volume ratio of core to sheath is much higher than the weight ratio of the net PEO–FITC to PEG–PLA, namely the observed core is actually a concentrated dispersion of PEO–FITC in PEG–PLA and sheath consists of pure PEG–PLA. (2) The emulsion droplets are stretched into an elliptical shape in the fiber direction for the fast evaporation speed of chloroform than that of water. Thus, the viscosity of the PEG–PLA/chloroform matrix increases more rapidly than that of the PEO–FITC/water droplets, and the viscosity of the outer layer of the fiber increases more rapidly than that of the deeper layer, resulting in the viscosity difference between the

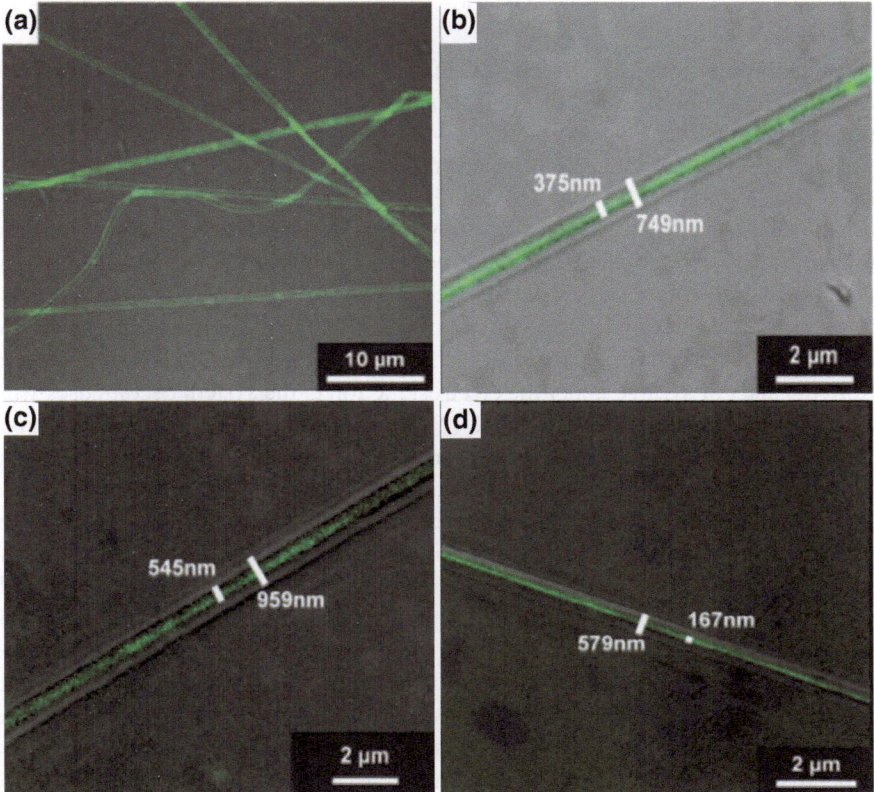

Fig. 3.32 Confocal laser scanning microscope (*CLSM*) images of the core–sheath-structured nanofibers prepared from W/O emulsions. The concentrations of core polymer (PEO–FITC) in the aqueous phase were as follows: **a, b** 45 mg/mL, **c** 58.5 mg/mL, and **d** 345.5 mg/mL. Reproduced from Ref. [52] by permission from John Wiley & Sons Ltd

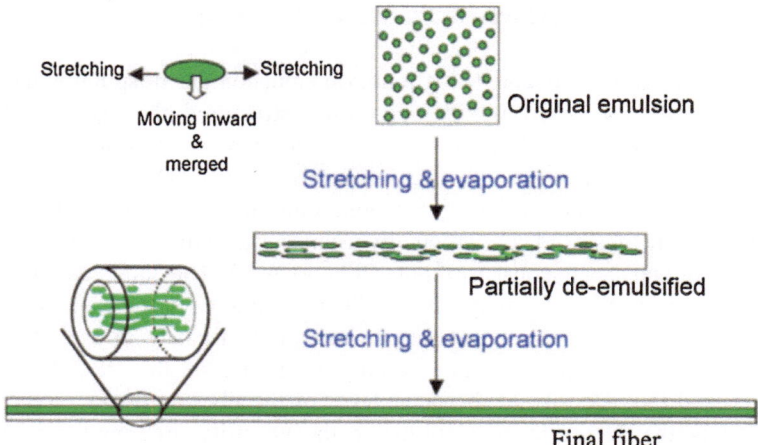

Fig. 3.33 Schematic mechanism for the formation of core–shell composite fibers during the emulsion electrospinning. Reproduced from Ref. [52] by permission from John Wiley & Sons Ltd

elliptical droplets and their matrix. (3) Inward movement of the emulsion droplets emerges, resulting in their emergence to some extent. In the extreme case, such phenomena may be quite thorough.

Recently, a novel interesting paper on fabricating poly (ethylene oxide)/chitosan oligosaccharide core–shell structures (Fig. 3.34) has been demonstrated by Zhang and Nie [53]. Moreover, they also systematically investigated the transformation of complex internal structures of poly (ethylene oxide)/chitosan oligosaccharide electrospun structures (Fig. 3.34). They suggested that the transformation of such complex structures is caused by both the immiscibility of the two polymers and phase separations: (1) As the polymer solution jets were ejected from the apex of syringe under the electrostatic force, the system decomposes with the continuous evaporation of solvent. (2) The location of the system composition, in the stable (unstable) phase region, determines the phase morphologies and structure within the fibers. (3) In concentrated solution (volume fraction of PEO is high), the quenching path of the system intersects the bimodal curve. At this time, phase separation of the system occurs in the metastable region bound by binodal and spinodal curves, in which the solution is separated into two phases; PEO-rich region and CS-rich region. (4) The nucleation of PEO takes place in the PEO-rich region. (5) The nucleation and growth process of the blend system might be intersected by the spinodal decomposition. Once it happens, the blending system separates sequentially with the mechanism of nucleation and growth process and spinodal decomposition, resulting in the structures containing both core–shell structure and internal phase-separated structures. From this mechanism, it can be found that the sequential phase transition with the mechanism of nucleation and growth process and spinodal decomposition is the important key for the structure transformation (Fig. 3.35).

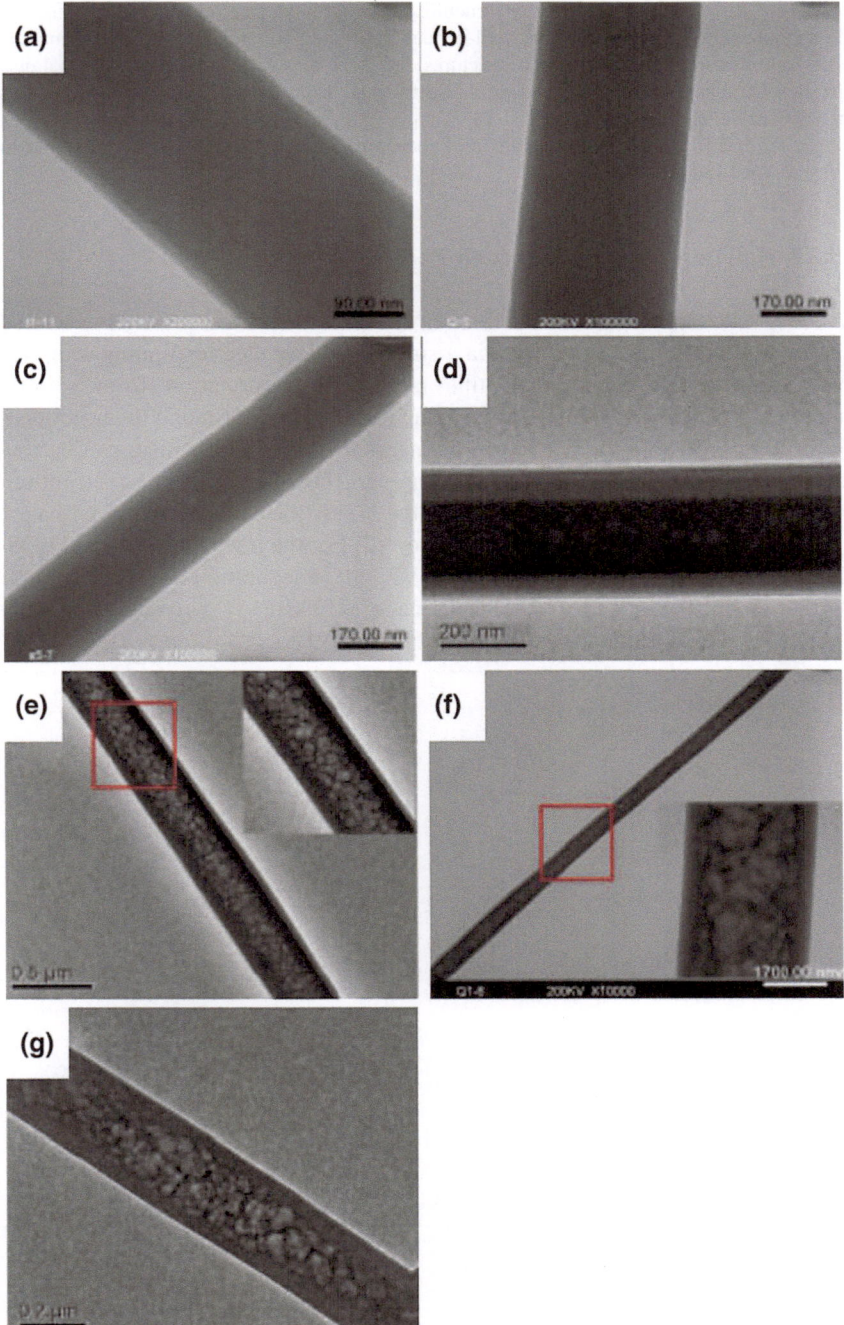

◀**Fig. 3.34** TEM images of the internal structures of poly (ethylene oxide)/chitosan oligosaccharide nanofibers. **a–c** Nanofibers with well-defined core–shell structures. **d, e** Transformation of core–shell to internal phase-separated structures. **f, g** Nanofiber with well-defined internal phase-separated structures. Those fibers were electrospun from poly (ethylene oxide)/chitosan oligosaccharide blend solution with $V_{poly\ (ethylene\ oxide)}/V_{chitosan\ oligosaccharide}$ ratios of **a** 2:1, **b** 1:1, **c** 1:2.5, **d** 1:3, **e** 1:3.5, **f** 1:4, and **g** 1:6. Reproduced from Ref. [53] by permission from John Wiley & Sons Ltd

3.3.2.3 Hollow Structures Through Phase Separation

In 2007, Li et al. [54], for the first time, presented a simple method for the fabrication of polymer nanotubes via a single capillary electrospinning setup from PVP/TEOS/ethanol solution. In this method, transformation from solidified nanofibers to hollow nanotube can be clearly seen (Fig. 3.36), and the introduction of TEOS in PVP solution is to form the nanotubes. The mechanism can be found as follows: (1) The ethanol evaporation is very fast, resulting in a concentration gradient of ethanol along the diameter of the fibers (the concentration of ethanol is highest in the center but lowest at the edge). (2) For the high solubility of TEOS in ethanol and the incompatibility against PVP, phase separation occurs, resulting in TEOS rich in center and PVP rich in edge (TEOS/PVP core-shell structures). (3) Finally, TEOS can evaporated from the wall of the PVP.

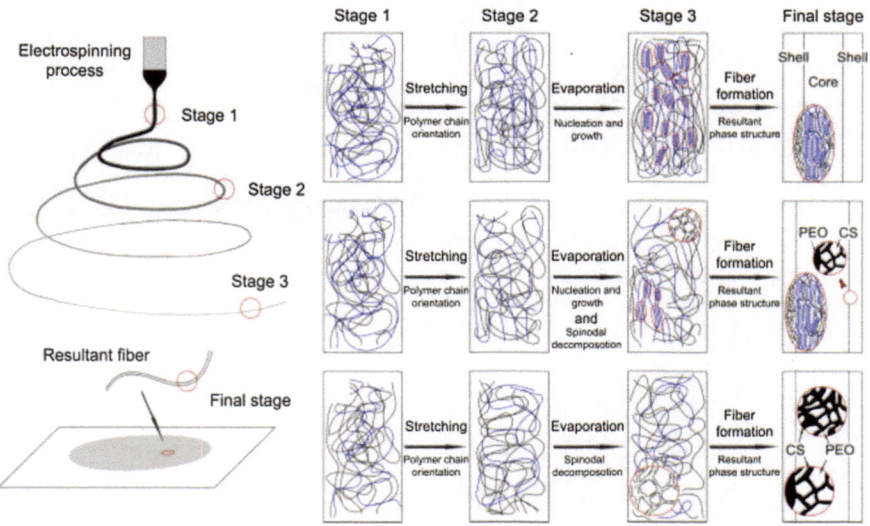

Fig. 3.35 Evolution of complex internal structures and structure transformation: **a** evolution of core–shell structures; **b** transformation of core–shell to internal phase-separated structure; **c** evolution of internal phase-separated structure. Reproduced from Ref. [53] by permission from John Wiley & Sons Ltd

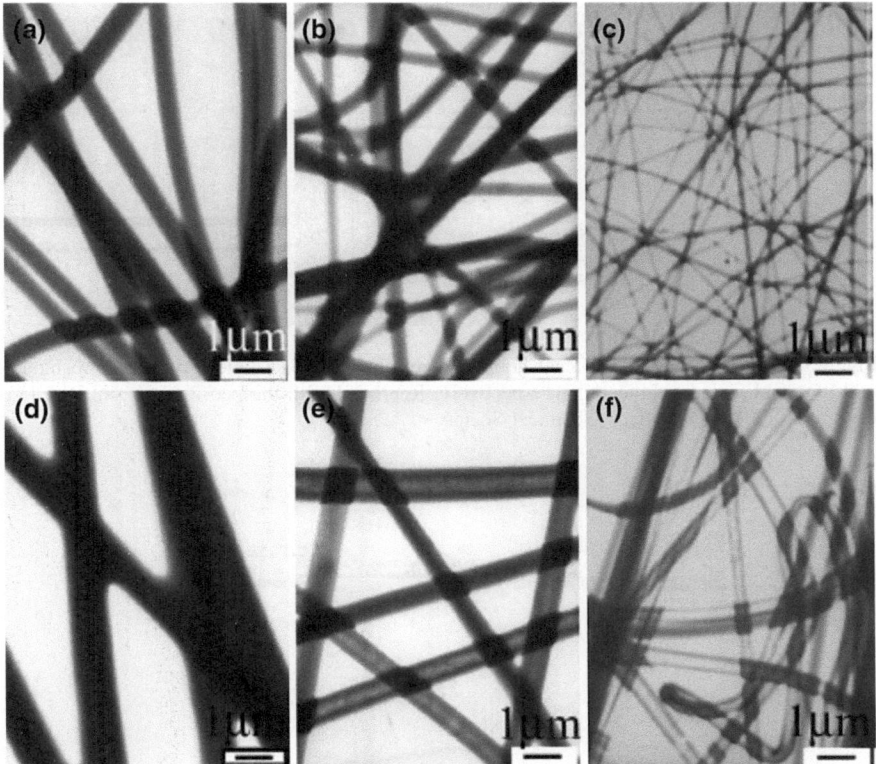

Fig. 3.36 TEM images of PVP nanofibers and PVP/TEOS nanotubes electrospun from 3 mL S1 added by 0 (**a**), 2 (**b**), and 4 mL (**c**) ethanol, and 0 (**d**), 2 (**e**), and 4 mL (**f**) ethanol with 0.3 mL TEOS, respectively. The introduction of TEOS in PVP solution leads to the formation of nanotubes. Reprinted with the permission from Ref. [54]. Copyright 2007 by American Chemical Society

3.3.2.4 Hollow Structures Through Emulsion Electrospinning

In 2009, Zhang et al. [55] obtained ZnO hollow nanofibers by electrospinning PAN/PVP–zinc acetate composites in DMF solution with single spinneret. Figure 3.37 displays the TEM images of the final ZnO hollow nanofibers with low and high magnifications.

In their experiment, incompatibility of the two polymers (PAN and PVP) plays an important role in the formation of hollow structures: (1) The incompatibility of the two polymers caused a dispersed phase and a continuous phase in the mixture polymer solution. (2) The mixture solution was metastable and decomposed into a stable emulsion owing to the high viscosity of PAN solution (0.035 Pa·s), which is nearly 5 times higher than that of the PVP/zinc acetate solution (0.0007 Pa·s). (3) Within the electrospinning process, the mixture solution will keep the phase separation rate with the dispersed phase (inner flow layer) of PAN and continuous phase (outer flow layer) of PVP/zinc acetate composites (Fig. 3.38).

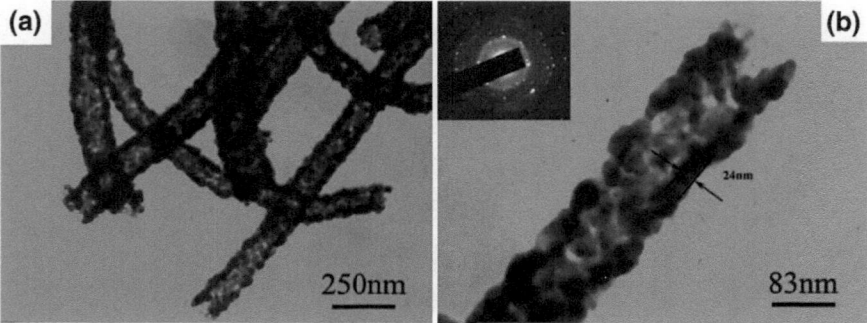

Fig. 3.37 TEM images of the ZnO hollow fibers obtained via single-syringe electrospinning with low and high magnifications, respectively. Reprinted with the permission from Ref. [55]. Copyright 2009 by American Chemical Society

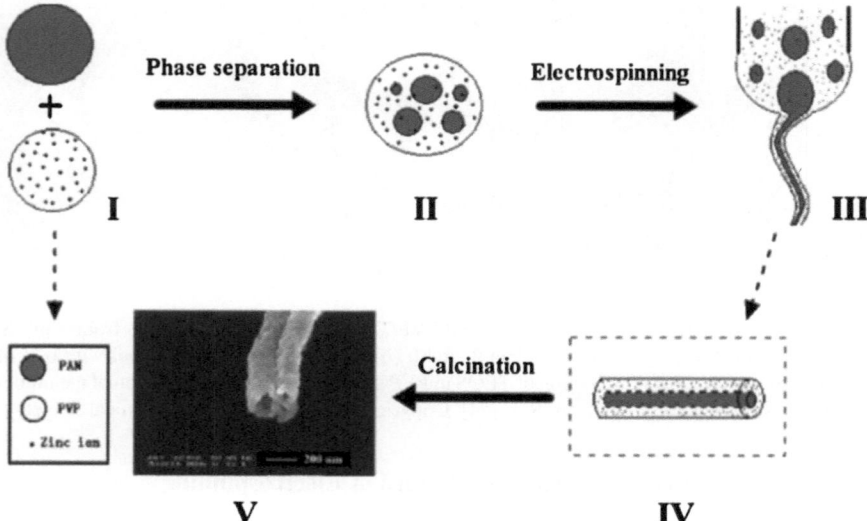

Fig. 3.38 Schematic illustration of the formation of hollow ZnO structures. Reprinted with the permission from Ref. [55]. Copyright 2009 by American Chemical Society

3.3.2.5 Hollow and Complex Structures Through Calcination

Recently, Guan and his groups [56, 57] successfully fabricated and solidified (Fig. 3.39a), hollow (Fig. 3.39b), and complex (fiber-in-tube: Fig. 3.39c; tube-in-tube: Fig. 3.39d) fibers through special calcination (a nonequilibrium heat treatment). It is important to note that this method is general in fabricating ceramic hollow and complex fibers.

They suggested that the formation of those complex structures is depended on the different heating rates (R):

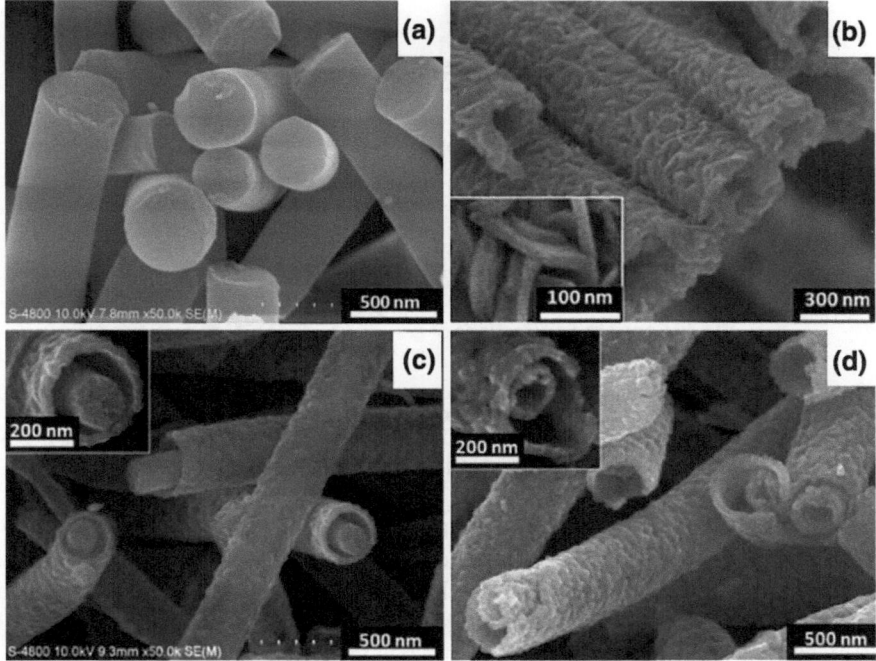

Fig. 3.39 SEM images of final Υ-Fe$_2$O$_3$ products (**a**), solidified fibers and complex structures containing hollow (**b**), fiber-in-tube (**c**), and tube-in-tube (**d**), respectively. The heating rates of (**a**), (**b**), (**c**), and (**d**) are 1, 50, 100 and 250 °C/min, respectively. Reprinted with the permission from Ref. [56]. Copyright 2010 by American Chemical Society

1. As the R is low, the electrospun precursor fibers can be homogeneously heated from the surface to the center and thus shrunk into solidified structures.
2. Increasing the R, a dense rigid shell will be first generated on the outer surface owing to the existence of large temperature gradient (ΔT) along the radial direction. At the same time, the dense rigid shell can prevent further contraction of the outer diameter, while the inner viscoelastic gel will continue to undergo intensive shrinkage for the loss of organic components in the subsequent calcination. In this case, it is imaginable that the interface layer between the inner gel and dense shell simultaneously receives two forces: the cohesive force (σ_{co}) from the inner gel and the adhesive force (σ_{ad}) with a direction opposite to that of the former. (1) If $\sigma_{co} \leq \sigma_{ad}$, the inner gel will shrink outward to the preformed dense shell along with the loss of organic components, resulting in hollow structures (R is 50 °C/min). (2) As the R is 100 °C/min, ΔT has been enlarged and σ_{ad} has been reduced, resulting in $\sigma_{co} \geq \sigma_{ad}$. Therefore, inner gel fiber will contract inward and separate from the dense shell during the calcination process, resulting in the fiber-in-tube structures. (3) If R is more higher (250 °C/min), ΔT within the inner gel fiber will be still strong enough; thus, dense shell will form on the outer surface of the inner gel, resulting in tube-in-tube structures (Fig. 3.40).

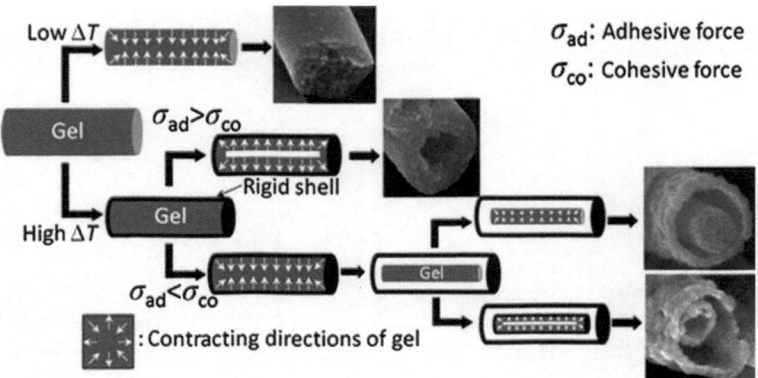

Fig. 3.40 Schematic diagram illustrates the formation of solid fiber (**a**), hollow fibers (**b**), fiber-in-tube (**c**), and tube-in-tube (**d**) nanostructures through different heating rate (R). Reprinted with the permission from Ref. [56]. Copyright 2010 by American Chemical Society

Inspired by this method, some groups successively obtained ceramic hollow fibers through calcination technique. For example, Xiang et al. [58] fabricated hollow CuO and Cu nanofibers with the heating rate of 6.7 °C/min (Fig. 3.41a and b). Kong et al. [59] fabricated hollow SnO_2 structures with the heating rate of 10 °C/min (Fig. 3.41c). Dong et al. [60] obtained Cr^{3+}-doped $ZnAl_2O_4$ hollow structures (Fig. 3.41d) with the heating rate of 3.3 °C/min.

3.3.2.6 Hollow or Open Cell Structures Through Compressed Carbon Dioxide

Recently, Liu et al. [61] combined the electrospinning and PCA (precipitation with a compressed fluid antisolvent) process (ES-PCA) to produce polymeric fibers with open cell or hollow core morphologies (Fig. 3.42). Carbon dioxide is used as the compressed gas of choice. In this chapter, the authors had investigated the impact of CO_2 pressure (density) on the fiber morphologies from different polymer/different solvent system (Fig. 3.43).

A mechanism has been established in this chapter to elucidate the formation of hollow or open cell structures: (1) CO_2 extracts the polymer solvent on the surface of the jet and makes the solid skin on the outer surface of jets form very rapidly. (2) Along with the CO_2 diffuse into the jets, the polymer-rich and polymer-lean phases start through the nucleation and growth mechanism. (3) The stretching of the jet promotes the coalescence of the CO_2-rich bubbles which displace the polymer-rich network from the center of the jet, ultimately resulting in the formation of hollow fiber (Fig. 3.44).

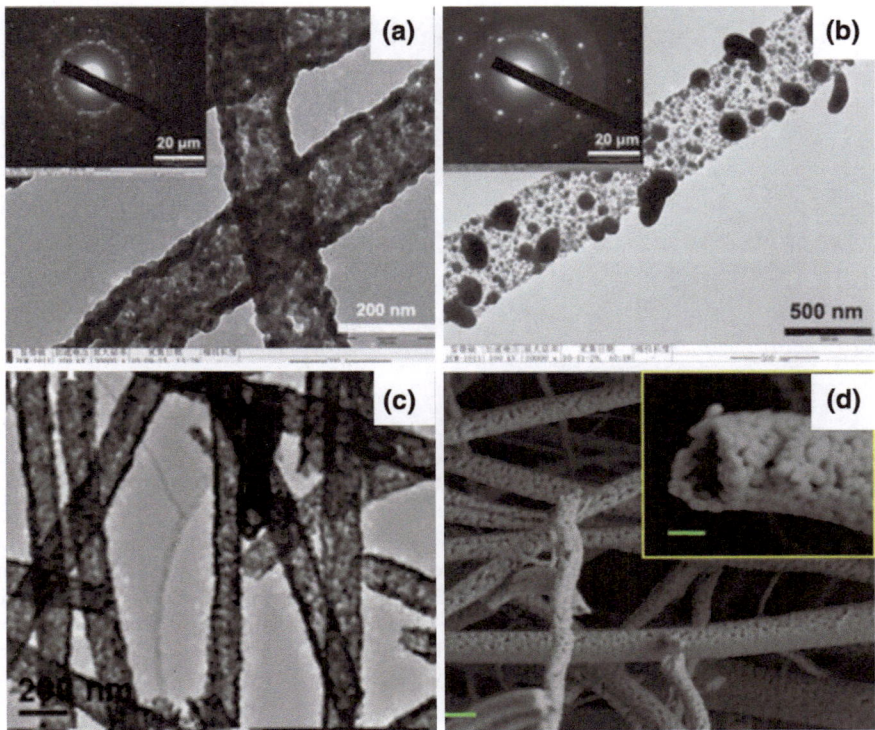

Fig. 3.41 Hollow structures obtained from single-spinneret electrospinning setup through the control of heating treatment. The heating rates of (**a** and **b**), (**c**), and (**d**) are 6.7, 10 and 3.3 °C/min, respectively. **a** and **b** Reprinted with the permission from Ref. [58] by permission from The Royal Society of Chemistry. **c** Reproduced from Ref. [59] by permission from The Royal Society of Chemistry; (**d**) Reproduced from Ref. [60] by permission from The Royal Society of Chemistry

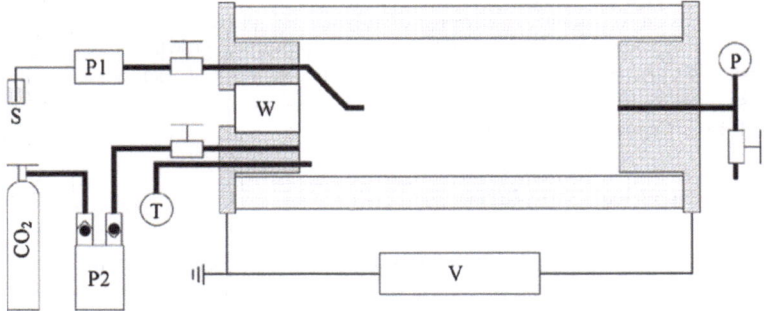

Fig. 3.42 Schematic diagram of the high-press, electrospinning setup. *P* is pressure gauge; *P1* and *P2* are solvent and carbon dioxide pumps; *S* is the polymer solution storage; *T* is temperature sensor; *W* and *V* are sapphire window and voltage supply. Reprinted with permission from Ref. [61]. Copyright 2010 by Elsevier

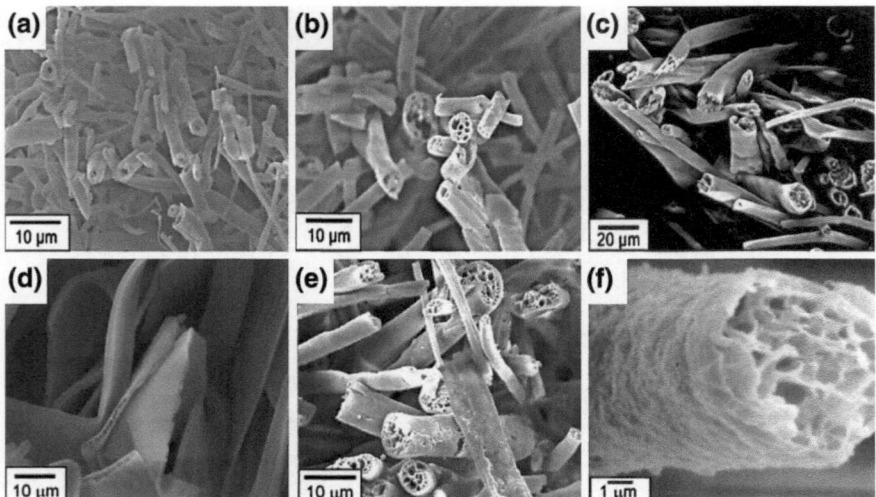

Fig. 3.43 SEM image of impart of CO_2 pressure on fiber morphologies. (A, B, and C) Views of 6.5 wt. % PVP dichloromethane into CO_2 at room temperature (22 °C) and 35 bar (**a**), 42 bar (**b**), and 50 bar (**c**), respectively. **d** View of PVP–ethanol solution into CO_2 at room temperature (22 °C) and 56 bar. **e** View of 5 wt. % PVP–ethanol solution into CO_2 at room temperature (22 °C) and 65 bar. **f** View of 15 wt. % PVDF–DMF solution into CO_2 at 42 °C and 90 bar. Reprinted with permission from Ref. [61]. Copyright 2010 by Elsevier

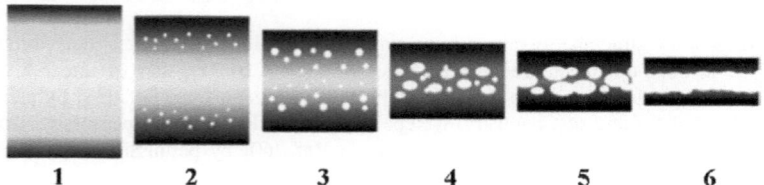

Fig. 3.44 Schematic illustration of hollow and open-cell structures from the electrospinning PCA process. *1* skin formation; *2* nucleation of polymer-rich phase (*darker region*), solvent-rich phase (*lighter regions*), and CO_2-rich phase (*white bubbles*); *3* growth of CO_2-rich phase; *4* coalescence of CO_2-rich phase bubbles; *5* further coalescence of CO_2-rich phase bubbles (if the process stops, an open cell structures will form); *6* hollow structures. Reprinted with permission from Ref. [61]. Copyright 2010 by Elsevier

3.3.2.7 Hierarchical Structures

In 2006, Ostermann et al. [62] successfully constructed hierarchical structures based on electrospinning fibers: V_2O_5 nanorods standing on TiO_2 electrospinning nanofibers through the calcination. Figure 3.46 shows the evolution of the V_2O_5-TiO_2 nanofibers calcinated at 475 °C with different calcination periods of time. A plausible mechanism on such hierarchical structures has been advanced: (1) At the nucleation stage of V_2O_5 nanorods, during the calcination, the presence of titania networks in the composite fibers or low temperature favoured the formation of

Fig. 3.45 SEM images on the evolution of the V_2O_5-TiO_2 nanofibers calcinated at 475 °C with different calcination periods of time. Reprinted with the permission from Ref. [62]. Copyright 2006 by American Chemical Society

Fig. 3.46 SEM images of the hierarchical brushy TiO_2 film synthesized by combining the electrospinning and hydrothermal route with low and high magnifications, respectively. Reproduced from Ref. [63] by permission from The Royal Society of Chemistry

small nuclei, resulting in the formation of thin and long nanorods. (2) The spatial confinement of vanadia by titania can force some of the V_2O_5 nanocrystal to grow toward the outside of the fibers to generate nanorods on the fibers, while some were trapped within the fibers. (3) The presence of titania can influence the crystallization and diffusion rates of V_2O_5 and further promote the growth of V_2O_5 nanorods (Fig. 3.45).

In 2008, Sun et al. [63] combined electrospinning and hydrothermal reaction method and successfully fabricated self-supporting hierarchical nanostructured anatase/rutile composite TiO_2 film (Fig. 3.46).

In contrast to the previous method for hierarchical structures (V_2O_5 nanorods standing on TiO_2 electrospinning nanofibers), outstanding advantages of this method lie not only in the ability to fabricate hierarchical nanostructured films with self-supporting properties, but also in a general method for the hierarchical nano-structured films. Driven by this method, a variety of hierarchical structured fibers have been realized including copper (II) phthalocyanine on TiO_2 hierarchical

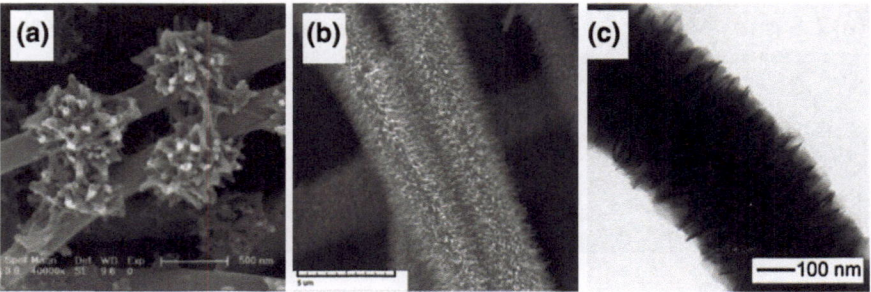

Fig. 3.47 SEM images of different hierarchical fibers by combining the electrospinning and hydrothermal technique. **a** Reprinted with the permission from Ref. [64]. Copyright 2011 by American Chemical Society. **b** Reproduced from Ref. [65] by permission from The Royal Society of Chemistry. **c** Reproduced from Ref. [66] by permission from The Royal Society of Chemistry

nanofibers (Fig. 3.47a) [64], ZnO/SiO$_2$ hierarchical nanofibers (Fig. 3.47b) [65], K-birnessite/TiO$_2$ hierarchical nanofibers (Fig. 3.47d) [66].

Recently, a novel method (wetting porous templates) has been combined with electrospinning technique to construct novel hierarchical fiber by Chen and groups [67]. Figure 3.48 shows the whole and simple process. The mechanism lies in the thermal annealing above the glass transition temperature of polymer nanofibers. At this time, the wetting of polymer chains within the polymer nanofiber will have enough mobility to move into the template. Finally, controllable hierarchical structures can be constructed on the outer surface of electrospun polymer fibers (Fig. 3.49).

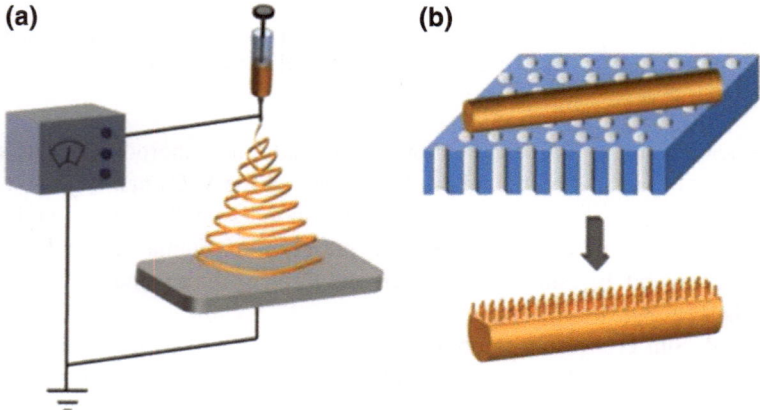

Fig. 3.48 Schematic illustration of the whole process to make hierarchical polymer structures based on electrospinning and wetting of porous templates. **a** The setup of electrospinning. **b** AAO templates on the top of polymer nanofibers. During the thermal annealing (temperature higher than the glass transition temperatures of polymer), the wetting polymer chains can enter into the template, resulting in nanorods on the top of the electrospun polymer fibers by removing the templates via selective etching solution such as weak base (NaOH). Reprinted with the permission from Ref. [67]. Copyright 2012 by American Chemical Society

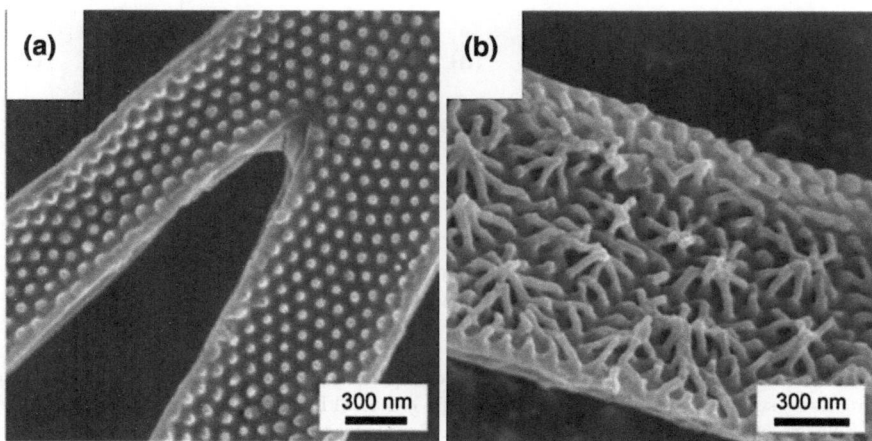

Fig. 3.49 SEM images of the hierarchical fibers containing nanorods on polymer fibers (**a**) and collapse nanorods on polymer fibers (**b**), respectively. Reprinted with the permission from Ref. [67]. Copyright 2012 by American Chemical Society

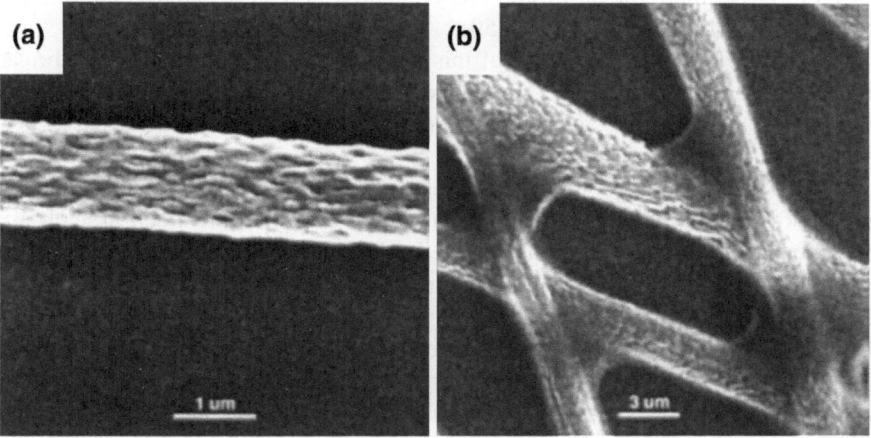

Fig. 3.50 SEM images of the interior porous structured fibers. **a** Residual PLA fiber after selective removal of PVP by water. **b** Residual PVP fibers after selective removal of PLA by annealing. Reproduced from Ref. [68] by permission from John Wiley & Sons Ltd

3.3.2.8 Interior Structure Within the Fiber Interior Pores Through the Polymer Blends

In 2001, Wendorff and groups [68] firstly build interior porous structures based on electrospun fibers through polymer blends with polylactide (PLA) and polyvinylpyrrolidone (PVP) as model. In this system, different polymeric phase separations, solubilities, and the decomposition temperature are the key for porous structures within the fibers. (1) Different polymeric phase separations form the

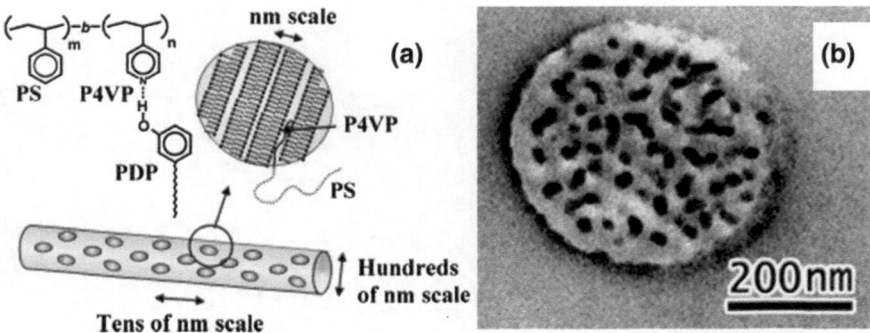

Fig. 3.51 a Schematic illustration of the electrospun fibers containing the internal periodic structures (self-assembled region of P4VP (PDP)$_1$). **b** TEM image of cross-sectional structure of the electrospun fiber after the cleaving of PDP. Reproduced from Ref. [69] by permission from John Wiley & Sons Ltd

PVP-rich and PLA-rich regions within the fibers. (2) The lower decomposition temperature of PLA makes the removal of PLA easy. (3) The good water solubility of PVP makes the removal of PVP easy. Finally, porous structures can be easily obtained in polymer fibers (Fig. 3.50).

It is important to emphasize that this chapter is the pioneer work to fabricate interior porous structured polymeric fibers through polymer blends. Inspired by this work, a number of interior porous structured polymeric fibers have been fabricated.

In 2005, Ruotsalainen et al. [69] designed a very interesting route for internal periodic structures within the electrospun fibers by hierarchical self-assembly of polymeric comb-shaped supramolecules. In this work, polystyrene-block-poly (4-vinylpyridine) (PS-b-P4VP) with 3-n-pentadecylphenol (PDP) was taken as model in which PDP can connect to the block of P4VP to form P4VP (PDP)$_1$ through hydrogen bond and tolerate the harsh electrospinning condition. Thus, internal periodic structures can be formed owing to the self-assembled region of P4VP (PDP)$_1$ (Fig. 3.51a). After electrospinning, the hydrogen-bonded PDP molecules can be cleaved, resulting in the internal periodic pores left in whole electrospun PS-b-P4VP polymeric fibers (Fig. 3.51b). In this study, the choice of PS-b-P4VP lies in the periodic microphase-separated structures that can be formed within the PS-b-P4VP membranes, which can provide periodic self-assembled regions of P4VP (PDP)$_1$ within the whole polymeric fibers.

3.3.2.9 Interior Pores Within the Fibers Through Phase Separation

In 2006, Xia and groups [70] constructed highly porous fibers by using cryogenic liquid as the collector (Fig. 3.52a). In this method, cryogenic liquid has been used as the collector; thus, those unevaporated solvent will be frozen within the polymer fiber, resulting in the solvent-rich (*solvent-rich means the polymer in poor region*)

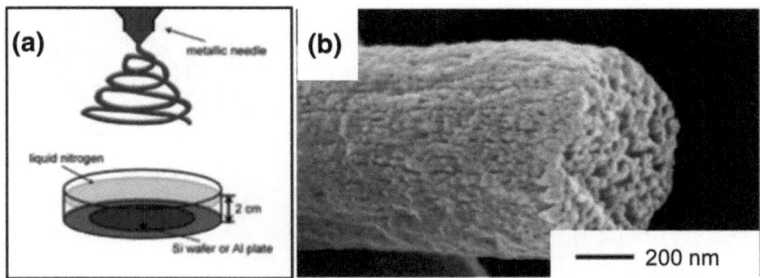

Fig. 3.52 **a** Schematic illustration of the whole electrospinning setup. **b** Porous structures within the whole PS nanofibers. Reprinted with the permission from Ref. [70]. Copyright 2006 by American Chemical Society

and solvent-poor regions. After the vacuum, porous structures can be easily obtained within the whole fibers (Fig. 3.52b).

3.3.2.10 Interior Complex Structure Within the Fiber

In 1999, Fong and Reneker [71] investigated the structures of electrospun styrene–butadiene–styrene triblock copolymer by staining with osmium tetroxide. They firstly found that separated phases of styrene and butadiene blocks can be observed within the fibers (Fig. 3.53). Although the interior complex structure within the fiber, caused by the separated phases, cannot be seen clearly then, this chapter does show

Fig. 3.53 TEM image of the electrospun styrene–butadiene–styrene triblock copolymer fiber. Reproduced from Ref. [71] by permission from John Wiley & Sons Ltd

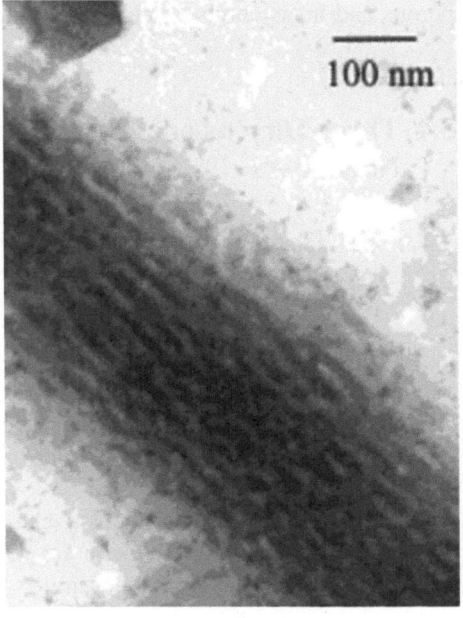

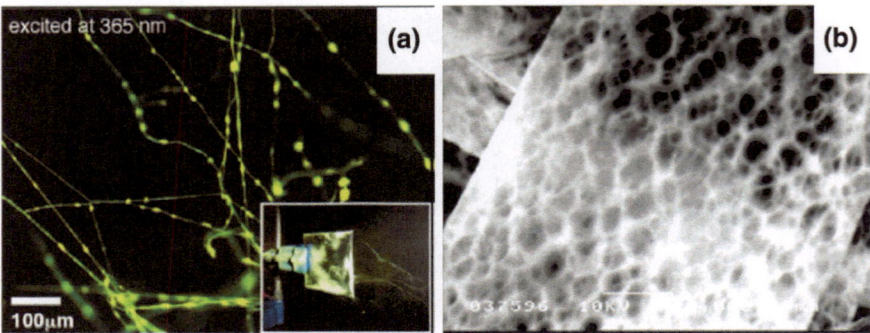

Fig. 3.54 **a** Fluorescent optical microscope images (insert: photograph of the nonwoven mat collected on ITO-PEN). **b** SEM image of the electrospun PTMSDPA fibers. Reprinted with the permission from Ref. [72]. Copyright 2006 by American Chemical Society

that there exists interior complex structures within the fibers by electrospinning of block copolymers. Follow this work, a lot of exciting interior complex structures have been found within the fiber via co-electrospinning, as mentioned in Fig. 3.29.

3.3.2.11 Nanoporous and Honeycomb-Structured Network Fiber

In 2006, Kwak et al. [72] demonstrated a novel structured fibers containing porous and honeycomb-structured network through the fiber with poly [1-(trimethylsilyl) phenyl-2-phenylacetylene] (PTMSDPA) as model (Fig. 3.54). However, the formation mechanism is not clear; the author suggests that the formation of such porous and honeycomb structures was caused by the atmospheric humidity.

3.4 Three-Dimensional Structures Formed by Electrospun Nanofibers

3.4.1 3D Honeycomb Structures Formed by Electrospinning Nanofibers

In 2006, Thandavamoorthy et al. [73] firstly observed that during the process of polyurethane (PU) electrospinning, 3D honeycomb structures could obtained through the self-assembly of PU nanofibers (Fig. 3.55). Inspired by this work, in 2011, Yan et al. [74] developed a series of 3D honeycomb structures with different polymers (Fig. 3.56).

The self-assembling mechanism can be explained as follows: The surface tension and electrostatic repulsion are the driving forces of the self-assembly of electrospun fibers; the surface tension is to make the nanofibers stick and merge

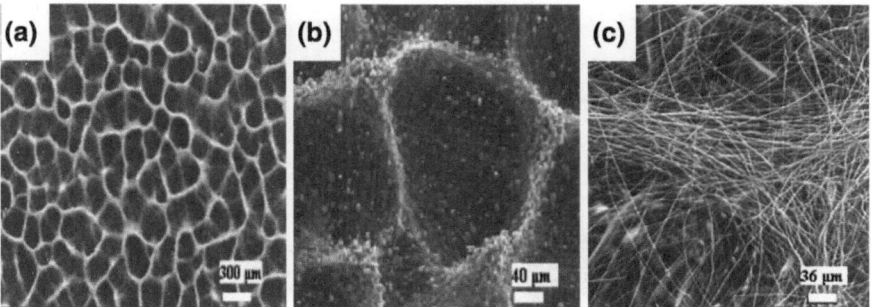

Fig. 3.55 SEM images of the 3D honeycomb structures by self-assembled PU electrospun with different magnifications. Reproduced from Ref. [73] by permission from John Wiley & Sons Ltd

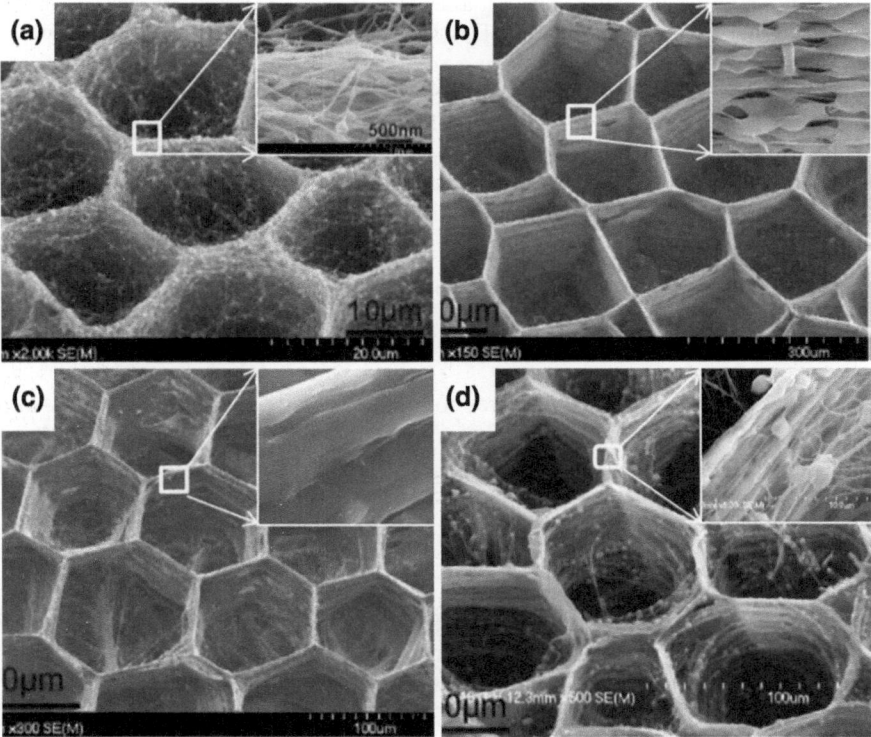

Fig. 3.56 SEM images of the 3D honeycomb structures with different polymers. **a** 3D honeycomb structures with 3 wt. % PAN from DMF solution at the voltage of 22 kV. **b** 3D honeycomb structures with 6 wt. % PVA from aqueous solution at the voltage of 22 kV. **c** 3D honeycomb structures with 16 wt. % PEO at the voltage of 22 kV. **d** 3D honeycomb structures with 16 wt. % PEO at the voltage of 19 kV. Reprinted with the permission from Ref. [74]. Copyright 2011 by American Chemical Society

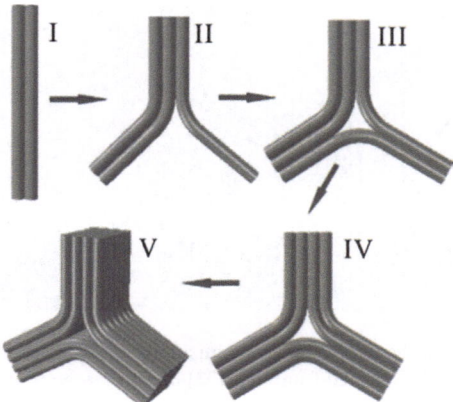

Fig. 3.57 Schematic illustration of the self-assembling mechanism of the electrospun nanofibers. Reprinted with the permission from Ref. [74]. Copyright 2011 by American Chemical Society

together when they contact each other, while the electrostatic repulsion resists this mergence and tries to push the fibers or the conglutinated fiber cluster away from each other when they approach; based on the competitive actions of those two forces, honeycomb structures can be formed (Fig. 3.57).

3.4.2 3D Complex Structures Formed by Electrospinning Nanofibers

Usually, other 3D complex structures formed by electrospinning nanofibers were obtained with the aid of fiber collection techniques. Recently, Zhang and Chang [75] summarized the other 3D complex structures formed by electrospinning nanofibers including 3D fibrous tubes (Fig. 3.58) and 3D multiple interconnected tubular structures (Fig. 3.59).

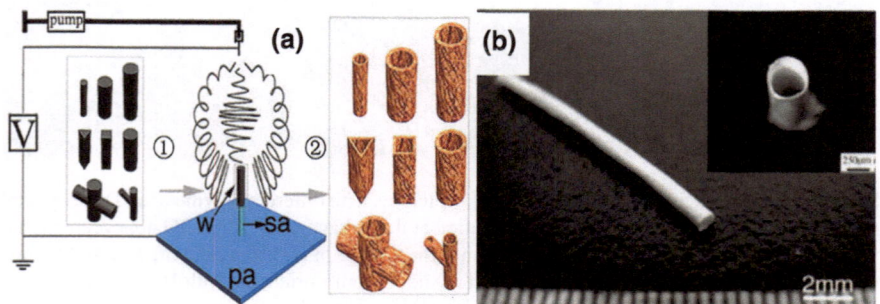

Fig. 3.58 a Schematic illustration of the process for fabrication of 3D fibrous tubes. **b** The relevant fibrous tubes (inset is the cross-sectional image). Reprinted with the permission from Ref. [75]. Copyright 2008 by American Chemical Society

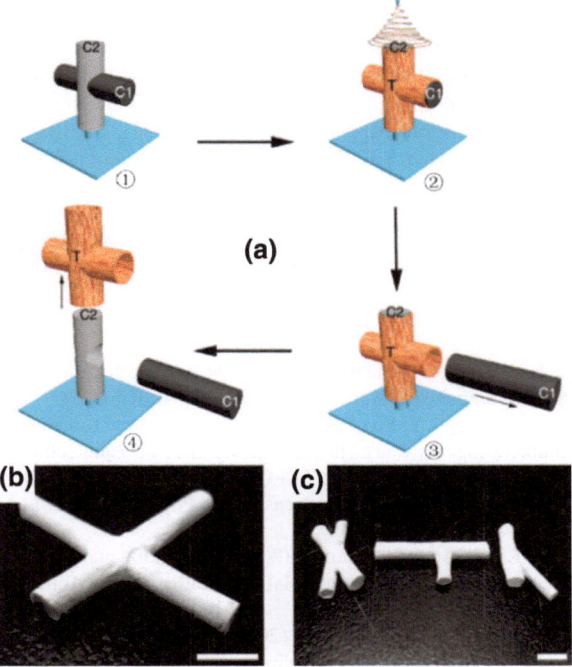

Fig. 3.59 a Schematic illustration of the process for 3D multiple interconnected tubular structures. b A crossing tube. c Tubes with various interconnected tubular structures. Reprinted with the permission from Ref. [75]. Copyright 2008 by American Chemical Society

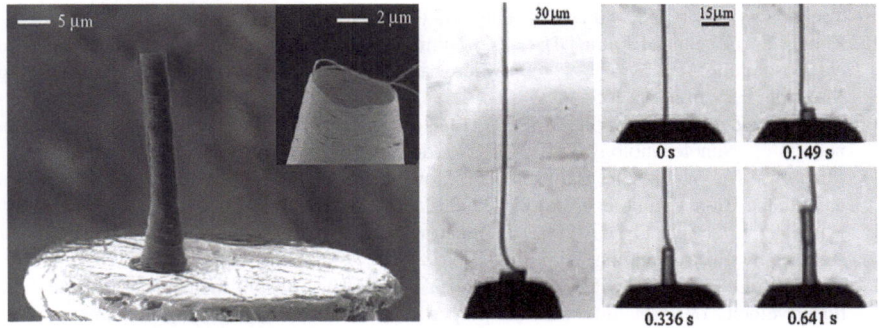

Fig. 3.60 Self-assembled electrospun hollow coiled structure built on the apex of a stainless steel conical tip. Reprinted with the permission from Ref. [76]. Copyright 2010 by American Chemical Society

Besides those 3D complex structures mentioned above, Kim et al. [76] also successfully fabricated nanopottery by taking advantage of the bending instability and manipulating the electric field to control the area of deposition as shown in Fig. 3.60.

References

1. Lee SW, Lee HJ, Choi JH, Koh WG, Myoung JM, Hur JH, Park JJ, Cho JH, Jeong U (2009) Periodic array of polyelectrolyte-gated organic transistors from electrospun poly(3-hexylthiophene) nanofibers. Nano Lett 10(1):347–351. doi:10.1021/nl903722z
2. Aviss KJ, Gough JE, Downes S (2010) ALIGNED electrospun polymer fibres for skeletal muscle regeneration. Eur Cells Mater 19:193–204
3. Campoy-Quiles M, Ishii Y, Sakai H, Murata H (2008) Highly polarized luminescence from aligned conjugated polymer electrospun nanofibers. Appl Phys Lett 92(21):213303–213305
4. Hohman MM, Shin M, Rutledge G, Brenner MP (2001) Electrospinning and electrically forced jets. I Stab Theory Phys Fluids 13(8):2201–2220
5. Sun DH, Chang C, Li S, Lin LW (2006) Near-field electrospinning. Nano Lett 6(4):839–842. doi:10.1021/nl0602701
6. Ruiz A, Vega E, Katiyar R, Valentin R (2007) Novel enabling wire bonding technology. In: 57th electronic components & technology conference, IEEE 2007, pp 458–462
7. Zheng GF, Dai YH, Wang LY, Sun DH (2007) Direct-write micro/nano-structure for flexible electronic manufacturing. In: 2007 7th IEEE conference on nanotechnology, vol 1–3
8. Chang C, Limkrailassiri K, Lin LW (2008) Continuous near-field electrospinning for large area deposition of orderly nanofiber patterns. Appl Phys Lett 93(12). doi:12311110.1063/1.2975834
9. Zhang Y, He XL, Li JP, Miao ZJ, Huang F (2008) Fabrication and ethanol-sensing properties of micro gas sensor based on electrospun SnO$_2$ nanofibers. Sens Actuators B 132(1):67–73. doi:10.1016/j.snb.2008.01.006
10. Zhang Y, Li JP, He XL, Gao XG, Jia J (2008) Deposition of tin oxide nanofibers on a microhotplate by a controlled electrospinning. Sens Lett 6(6):956–960. doi:10.1166/sl.2008.539
11. Wang HL, Zheng GF, Sun DH (2009) Simulation of nanofibers movement for near-field electrospinning. In: Wang X (ed) Micro and nano technology, vol 60–61. Advanced Materials Research pp 456–460
12. Zheng GF, Wang LY, Wang HL, Sun DH, Li WW, Lin LW (2009) Deposition characteristics of direct-write suspended micro/nano-structures. In: Wang X (ed) Micro and Nano Technology, vol 60-61. Advanced Materials Research, pp 439–444
13. Zhong YY, Zheng GF, Sun DH (2009) Numerical simulation of viscous jet for near-field electrospinning. In: Wang X (ed) Micro and nano technology, vol 60–61. Advanced Materials Research. pp 465–469
14. Theron A, Zussman E, Yarin AL (2001) Electrostatic field-assisted alignment of electrospun nanofibres. Nanotechnology 12(3):384–390. doi:10.1088/0957-4484/12/3/329
15. Matthews JA, Wnek GE, Simpson DG, Bowlin GL (2002) Electrospinning of collagen nanofibers. Biomacromolecules 3(2):232–238. doi:10.1021/bm015533u
16. Kameoka J, Craighead HG (2003) Fabrication of oriented polymeric nanofibers on planar surfaces by electrospinning. Appl Phys Lett 83(2):371–373. doi:10.1063/1.1592638
17. Kim KW, Lee KH, Khil MS, Ho YS, Kim HY (2004) The effect of molecular weight and the linear velocity of drum surface on the properties of electrospun poly(ethylene terephthalate) nonwovens. Fibers Polym 5(2):122–127. doi:10.1007/bf02902925
18. Mathew G, Hong JP, Rhee JM, Leo DJ, Nah C (2006) Preparation and anisotropic mechanical behavior of highly-oriented electrospun poly(butylene terephthalate) fibers. J Appl Polym Sci 101(3):2017–2021. doi:10.1002/app.23762
19. Kim K, Lee K, Khil M, Ho Y, Kim H (2004) The effect of molecular weight and the linear velocity of drum surface on the properties of electrospun poly(ethylene terephthalate) nonwovens. Fibers Polym 5(2):122–127. doi:10.1007/bf02902925
20. Katta P, Alessandro M, Ramsier RD, Chase GG (2004) Continuous electrospinning of aligned polymer nanofibers onto a wire drum collector. Nano Lett 4(11):2215–2218. doi:10.1021/nl0486158

21. Pan H, Li L, Hu L, Cui X (2006) Continuous aligned polymer fibers produced by a modified electrospinning method. Polymer 47(14):4901–4904. doi:10.1016/j.polymer.2006.05.012
22. Li D, Wang Y, Xia Y (2003) Electrospinning of polymeric and ceramic nanofibers as uniaxially aligned arrays. Nano Lett 3(8):1167–1171. doi:10.1021/nl0344256
23. Li D, Wang Y, Xia Y (2004) Electrospinning nanofibers as uniaxially aligned arrays and layer-by-layer stacked films. Adv Mater 16(4):361–366. doi:10.1002/adma.200306226
24. Teo WE, Ramakrishna S (2005) Electrospun fibre bundle made of aligned nanofibres over two fixed points. Nanotechnology 16(9):1878–1884. doi:10.1088/0957-4484/16/9/077
25. Dalton PD, Klee D, Möller M (2005) Electrospinning with dual collection rings. Polymer 46(3):611–614. doi:10.1016/j.polymer.2004.11.075
26. Khil M-S, Bhattarai SR, Kim H-Y, Kim S-Z, Lee K-H (2005) Novel fabricated matrix via electrospinning for tissue engineering. J Biomed Mater Res B Appl Biomater 72B(1):117–124. doi:10.1002/jbm.b.30122
27. Smit E, Büttner U, Sanderson RD (2005) Continuous yarns from electrospun fibers. Polymer 46(8):2419–2423. doi:10.1016/j.polymer.2005.02.002
28. Bognitzki M, Czado W, Frese T, Schaper A, Hellwig M, Steinhart M, Greiner A, Wendorff JH (2001) Nanostructured fibers via electrospinning. Adv Mater 13(1):70–72. doi:10.1002/1521-4095(200101)13:1<70:aid-adma70>3.0.co;2-h
29. Megelski S, Stephens JS, Chase DB, Rabolt JF (2002) Micro- and nanostructured surface morphology on electrospun polymer fibers. Macromolecules 35(22):8456–8466. doi:10.1021/ma020444a
30. Casper CL, Stephens JS, Tassi NG, Chase DB, Rabolt JF (2004) Controlling surface morphology of electrospun polystyrene fibers: effect of humidity and molecular weight in the electrospinning process. Macromolecules 37(2):573–578. doi:10.1021/ma0351975
31. Caneba GT, Soong DS (1985) Polymer membrane formation through the thermal-inversion process. 2. Mathematical modeling of membrane structure formation. Macromolecules 18(12):2545–2555. doi:10.1021/ma00154a032
32. Laxminarayan A, McGuire KS, Kim SS, Lloyd DR (1994) Effect of initial composition, phase separation temperature and polymer crystallization on the formation of microcellular structures via thermally induced phase separation. Polymer 35(14):3060–3068. doi:10.1016/0032-3861(94)90420-0
33. Xue Y, Wang H, Yu D, Feng L, Dai L, Wang X, Lin T (2009) Superhydrophobic electrospun POSS-PMMA copolymer fibres with highly ordered nanofibrillar and surface structures. Chem Commun 42:6418–6420
34. Huang C, Tang Y, Liu X, Sutti A, Ke Q, Mo X, Wang X, Morsi Y, Lin T (2011) Electrospinning of nanofibres with parallel line surface texture for improvement of nerve cell growth. Soft Matter 7(22):10812–10817
35. Sun ZC, Zussman E, Yarin AL, Wendorff JH, Greiner A (2003) Compound core-shell polymer nanofibers by co-electrospinning. Adv Mater 15(22):1929–1932. doi:10.1002/adma.200305136
36. Lam HL, Ye HH, Gogotsi Y, Ko FK (2004) Structure and properties of single-walled carbon nanotubes reinforced nanocomposite fibrils by co-electrospinning. Abs Pap Am Chem Soc 228:U375–U375
37. Li D, Babel A, Jenekhe SA, Xia YN (2004) Nanofibers of conjugated polymers prepared by electrospinning with a two-capillary spinneret. Adv Mater 16(22):2062–2066. doi:10.1002/adma.200400606
38. Dong CH, Yuan XY, He MY, Yao KD (2006) Preparation of PVA/PEI ultra-fine fibers and their composite membrane with PLA by electrospinning. J Biomater Sci Polym 17(6):631–643. doi:10.1163/156856206777346287
39. Reznik SN, Yarin AL, Zussman E, Bercovici L (2006) Evolution of a compound droplet attached to a core-shell nozzle under the action of a strong electric field. Phys Fluids 18(6). doi:0621010.1063/1.2206747

40. Zussman E, Yarin AL, Bazilevsky AV, Avrahami R, Feldman M (2006) Electrospun polyacrylonitrile/poly (methyl methacrylate)-derived turbostratic carbon micro-/nanotubes. Adv Mater 18(3):348–353. doi:10.1002/adma.200501153

41. Halaui R, Moldavsky A, Cohen Y, Semiat R, Zussman E (2011) Development of micro-scale hollow fiber ultrafiltration membranes. J Membr Sci 379(1–2):370–377. doi:10.1016/j.memsci.2011.06.011

42. Du P, Song L, Xiong J, Yuan Y, Wang L, Xi Z, Jin D, Chen J (2012) TiO2/Nb2O5 core–sheath nanofibers film: co-electrospinning fabrication and its application in dye-sensitized solar cells. Electrochem Commun 25:46–49. doi:10.1016/j.elecom.2012.09.013

43. Xin Y, Huang Z, Li W, Jiang Z, Tong Y, Wang C (2008) Core–sheath functional polymer nanofibers prepared by co-electrospinning. Eur Polymer J 44(4):1040–1045. doi:10.1016/j.eurpolymj.2008.01.034

44. Yang Z, Cao C, Liu F, Chen D, Jiao X (2010) Core–shell Li(Ni1/3Co1/3Mn1/3)O2/Li(Ni1/2Mn1/2)O2 fibers: synthesis, characterization and electrochemical properties. Solid State Ionics 181(15–16):678–683. doi:10.1016/j.ssi.2010.03.032

45. Li D, Xia YN (2004) Direct fabrication of composite and ceramic hollow nanofibers by electrospinning. Nano Lett 4(5):933–938. doi:10.1021/nl049590f

46. McCann JT, Marquez M, Xia YN (2006) Melt coaxial electrospinning: a versatile method for the encapsulation of solid materials and fabrication of phase change nanofibers. Nano Lett 6(12):2868–2872. doi:10.1021/nl0620839

47. Zhao Y, Cao XY, Jiang L (2007) Bio-mimic multichannel microtubes by a facile method. J Am Chem Soc 129(4):764–765. doi:10.1021/ja068165g

48. Chen H, Wang N, Di J, Zhao Y, Song Y, Jiang L (2010) Nanowire-in-microtube structured core/shell fibers via multifluidic coaxial electrospinning. Langmuir 26(13):11291–11296. doi:10.1021/la100611f

49. Liu Z, Sun DD, Guo P, Leckie JO (2007) An Efficient bicomponent TiO2/SnO2 nanofiber photocatalyst fabricated by electrospinning with a side-by-side dual spinneret method. Nano Lett 7(4):1081–1085. doi:10.1021/nl061898e

50. Ma ML, Krikorian V, Yu JH, Thomas EL, Rutledge GC (2006) Electrospun polymer nanofibers with internal periodic structure obtained by microphase separation of cylindrically confined block copolymers. Nano Lett 6(12):2969–2972. doi:10.1021/nl062311z

51. Li ZY, Huang HM, Wang C (2006) Electrostatic forces induce poly(vinyl alcohol)-protected copper nanoparticles to form copper/poly(vinyl alcohol) nanocables via electrospinning. Macromol Rapid Commun 27(2):152–155. doi:10.1002/marc.200500627

52. Xu XL, Zhuang XL, Chen XS, Wang XR, Yang LX, Jing XB (2006) Preparation of core-sheath composite nanofibers by emulsion electrospinning. Macromol Rapid Commun 27(19):1637–1642. doi:10.1002/marc.200600384

53. Zhang JF, Nie J (2012) Transformation of complex internal structures of poly(ethylene oxide)/chitosan oligosaccharide electrospun nanofibers. Polym Int 61(1):135–140. doi:10.1002/pi.3159

54. Li XH, Shao CL, Liu YC (2007) A simple method for controllable preparation of polymer nanotubes via a single capillary electrospinning. Langmuir 23(22):10920–10923. doi:10.1021/la701806f

55. Zhang Z, Li X, Wang C, Wei L, Liu Y, Shao C (2009) ZnO hollow nanofibers: fabrication from facile single capillary electrospinning and applications in gas sensors. J Phys Chem C 113(45):19397–19403. doi:10.1021/jp9070373

56. Mou F, J-g Guan, Shi W, Sun Z, Wang S (2010) Oriented contraction: a facile nonequilibrium heat-treatment approach for fabrication of maghemite fiber-in-tube and tube-in-tube nanostructures. Langmuir 26(19):15580–15585. doi:10.1021/la102830p

57. Mou FZ, Guan JG, Sun ZG, Fan XA, Tong GX (2010) In situ generated dense shell-engaged Ostwald ripening: a facile controlled-preparation for BaFe12O19 hierarchical hollow fiber arrays. J Solid State Chem 183(3):736–743. doi:10.1016/j.jssc.2010.01.016

58. Xiang HF, Long YH, Yu XL, Zhang XL, Zhao N, Xu J (2011) A novel and facile method to prepare porous hollow CuO and Cu nanofibers based on electrospinning. Cryst Eng Comm 13(15):4856–4860. doi:10.1039/c0ce00980f

59. Kong JH, Wong SY, Zhang Y, Tan HR, Li X, Lu XH (2011) One-dimensional carbon-SnO_2 and SnO_2 nanostructures via single-spinneret electrospinning: tunable morphology and the underlying mechanism. J Mater Chem 21(40):15928–15934. doi:10.1039/c1jm12492g

60. Dong GP, Xiao XD, Peng MY, Ma ZJ, Ye S, Chen DD, Qin HJ, Deng GL, Liang QM, Qiu JR (2012) Synthesis and optical properties of chromium-doped spinel hollow nanofibers by single-nozzle electrospinning. RSC Adv 2(7):2773–2782. doi:10.1039/c2ra00516f

61. Liu J, Shen ZH, Lee SH, Marquez M, McHugh MA (2010) Electrospinning in compressed carbon dioxide: hollow or open-cell fiber formation with a single nozzle configuration. J Supercrit Fluids 53(1–3):142–150. doi:10.1016/j.supflu.2010.02.016

62. Ostermann R, Li D, Yin YD, McCann JT, Xia YN (2006) V_2O_5 nanorods on TiO_2 nanofibers: a new class of hierarchical nanostructures enabled by electrospinning and calcination. Nano Lett 6(6):1297–1302. doi:10.1021/nl060928a

63. Sun CH, Wang NX, Zhou SY, Hu XJ, Chen P (2008) Preparation of self-supporting hierarchical nanostructured anatase/rutile composite TiO_2 film (pg 3293, 2008). Chem Commun 48:6612. doi:10.1039/b805072d

64. Zhang MY, Shao CL, Guo ZC, Zhang ZY, Mu JB, Cao TP, Liu YC (2011) Hierarchical nanostructures of copper(II) phthalocyanine on electrospun TiO_2 nanofibers: controllable solvothermal-fabrication and enhanced visible photocatalytic properties. ACS Appl Mater Interfaces 3(2):369–377. doi:10.1021/am100989a

65. Wang RY, Guo J, Chen D, Miao YE, Pan JS, Tjiu WW, Liu TX (2011) "Tube brush" like ZnO/SiO_2 hybrid to construct a flexible membrane with enhanced photocatalytic properties and recycling ability. J Mater Chem 21(48):19375–19380. doi:10.1039/c1jm13979g

66. Dai YQ, Lu XF, McKiernan M, Lee EP, Sun YM, Xia YN (2010) Hierarchical nanostructures of K-birnessite nanoplates on anatase nanofibers and their application for decoloration of dye solution. J Mater Chem 20(16):3157–3162. doi:10.1039/c000446d

67. Chen JT, Chen WL, Fan PW (2012) Hierarchical structures by wetting porous templates with electrospun polymer fibers. ACS Macro Lett 1(1):41–46. doi:10.1021/mz200008e

68. Bognitzki M, Frese T, Steinhart M, Greiner A, Wendorff JH, Schaper A, Hellwig M (2001) Preparation of fibers with nanoscaled morphologies: electrospinning of polymer blends. Polym Eng Sci 41(6):982–989. doi:10.1002/pen.10799

69. Ruotsalainen T, Turku J, Heikkila P, Ruokolainen J, Nykanen A, Laitinen T, Torkkeli M, Serimaa R, ten Brinke G, Harlin A, Ikkala O (2005) Towards internal structuring of electrospun fibers by hierarchical self-assembly of polymeric comb-shaped supramolecules. Adv Mater 17(8):1048–1052. doi:10.1002/adma.200401530

70. McCann JT, Marquez M, Xia YN (2006) Highly porous fibers by electrospinning into a cryogenic liquid. J Am Chem Soc 128(5):1436–1437. doi:10.1021/ja056810y

71. Fong H, Reneker DH (1999) Elastomeric nanofibers of styrene-butadiene-styrene triblock copolymer. J Polym Sci Part B-Polym Phys 37(24):3488–3493. doi:10.1002/(sici)1099-0488(19991215)37:24<3488:aid-polb9>3.0.co;2-m

72. Kwak G, Fukao S, Fujiki M, Sakaguchi T, Masuda T (2006) Nanoporous, honeycomb-structured network fibers spun from semiflexible, ultrahigh molecular weight, disubstituted aromatic polyacetylenes: superhierarchical structure and unique optical anisotropy. Chem Mater 18(23):5537–5542. doi:10.1021/cm061719u

73. Thandavamoorthy S, Gopinath N, Ramkumar SS (2006) Self-assembled honeycomb polyurethane nanofibers. J Appl Polym Sci 101(5):3121–3124. doi:10.1002/app.24333

74. Yan GD, Yu J, Qiu YJ, Yi XH, Lu J, Zhou XS, Bai XD (2011) Self-assembly of electrospun polymer nanofibers: a general phenomenon generating honeycomb-patterned nanofibrous structures. Langmuir 27(8):4285–4289. doi:10.1021/la1047936

75. Zhang DM, Chang J (2008) Electrospinning of three-dimensional nanofibrous tubes with controllable architectures. Nano Lett 8(10):3283–3287. doi:10.1021/nl801667s

76. Kim HY, Lee M, Park KJ, Kim S, Mahadevan L (2010) Nanopottery: coiling of electrospun polymer nanofibers. Nano Lett 10(6):2138–2140. doi:10.1021/nl100824d

Chapter 4
Applications of Electrospun Nanofibers

Abstract With the development of electrospinning, diverse materials have been successfully used in electrospinning for nanofibers including organic materials, organic/inorganic materials, and inorganic materials with the diameter range from microscale down to nanoscale. Additionally, variety of architectures based on electrospun fibers have been realized such as solidified fibers, porous (surface pores or interior pores) fibers, hollow fibers, core–shell fibers, hierarchical structured fibers. Those unique advantages of electrospinning afford the multi-functional properties for diverse applications, including nanofiber reinforcement, filtration, catalysis, electronic devices, lithium-ion battery, fuel cells, biomedical field, etc. In this chapter, we intend to provide an exposition of some special activities.

Keywords Electrospinning · Nanofiber reinforcement · Filtration · Catalysis · Electronic devices · Lithium-ion battery · Fuel cells · Biomedical field

In this chapter, we will present the applications of electrospun nanostructures in nanofiber reinforcement, filtration, catalysis, electronic devices, lithium-ion battery, fuel cells, and biomedical field, in turn. Within each of those applications, the basic research and the challenges will be presented.

4.1 Nanofiber Reinforcement

One of the most important applications based on traditional fibers at microscale, such as carbon fiber, glass fiber, and Kevlar fibers, is to act as fiber reinforcement in composite systems [1]. With the aid of those fiber reinforcements, the composite materials can exhibit superior structural properties including high modulus and strength to weight ratios. Prompted by those practical applications based on traditional fibers, electrospun nanofibers have been investigated in fiber reinforcement as well, driven by their better mechanical properties than bulk fiber [2].

Z. Li and C. Wang, *One-Dimensional Nanostructures*,
SpringerBriefs in Materials, DOI: 10.1007/978-3-642-36427-3_4,
© The Author(s) 2013

4.1.1 Basic Research on Nanofiber-Reinforced Composites

In 1999, Kim and Reneker [3] firstly investigated the nanofiber-reinforcing effect of the polybenzimidazole (PBI) nanofiber in epoxy matrix and rubber matrix, respectively. For epoxy composite: 8–32 plies of the fiber sheets were cut and folded to fit the compression mold, and then epoxy was impregnated with the fiber sheets to form resin-rich layer between adjacent fiber sheets. The cure was done in an elevated vacuum oven (150 °C for 3 h then followed by 200 °C for 2 h). Specimens 60 mm long, 30 mm wide and 3 mm thick were cut from the molded sheets; for rubber composite: Cut the nonwoven sheet of nanofibers into 0.5 cm^2. The sample was milled for times sufficient to disperse the fibers into the matrix. Sample sheets were cured by compression molding at 160 °C for 1 h. Fiber contents within the composites were determined by extracting the fibers from the uncured mixture with toluene.

Nanofiber-reinforcing effect for epoxy composite: Increasing content of electrospun fibers, Young's modulus (E), the fracture toughness (K_{Ic}), and the fracture energy (G_{Ic}) of the composites will be increased as shown in Table 4.1. Nanofiber-reinforcing effect for rubber composite: the Young's modulus was ten times and the tear strength was twice as large as that of the bulk rubber materials without fiber as shown in Table 4.2.

Bergshoef and Vansco [2] also investigated fiber-reinforcing effect of the Nylon-4/6 nanofiber nonwoven in diluted epoxy matrix. They found that both the stiffness and strength of the composite are significantly higher in contrast to the reference film (Table 4.3). In addition, the elongation at break was significantly higher for the reference sheet. Furthermore, the epoxy composite with nylon-4/6 nanofiber exhibited a characteristic transparency owing to the fiber sizes smaller than the wavelength of visible light.

Besides the stiffness and strength enforcement, scientists also tried to modify other mechanical behaviors of the composite with the aid of the electrospun

Table 4.1 The mechanical properties of the Fiber-epoxy composite

Fiber content (By weight wt. %)		E (Gpa)	K_{Ic} (Mpa M$_{1/2}$)	G_{Ic} (J/M^2)
0		4.25 ± 0.01	0.63	93.4
3	A	5.27 ± 0.01	0.72 ± 0.03	98.4 ± 0.05
	B		0.86 ± 0.03	140.3 ± 0.03
9	A	5.33 ± 0.02	0.92 ± 0.03	158.4 ± 0.10
	B		0.99 ± 0.05	183.9 ± 0.08
12	A	5.35 ± 0.03	0.96 ± 0.06	172.3 ± 0.11
	B		1.23 ± 0.08	282.3 ± 0.12
15	A	5.40 ± 0.03	1.11 ± 0.07	228.2 ± 0.15
	B		1.40 ± 0.08	363.0 ± 0.15

A denotes the winding direction composites, and *B* denotes the transverse direction composites. Reproduced from Ref. [3] by permission of John Wiley and Sons Ltd

Table 4.2 The mechanical properties of rubber composite

	Young's modulus (Mpa)	Tensile strength (Mpa)	Tear strength (kN/m)
Unfilled	1.8 ± 0.05	2.1 ± 0.05	9.1 ± 0.05
Fiber-filled rubber	19.6 ± 0.10	2.8 ± 0.10	17.4 ± 0.10

Reproduced from Ref. [3] by permission of John Wiley and Sons Ltd

Tables 4.3 Mechanical properties of nylon-4,6/epoxy composite films and epoxy reference films

	Young's moduli (Mpa)	Fracture stresses[a] (Mpa)	Fracture strains[a] (%)
Composite	91	2.4	3.0
Reference sample[b]	2.5	0.82	8.0

[a] Yielding occurs at $\sigma = 16$ Mpa and $\varepsilon = 15$ %. [b] Specimen thickness: 20 mm
σ: Fracture stresses; ε: Fracture strains. Reproduced from Ref. [2] by permission of John Wiley and Sons Ltd

nanofibers. Prof. Dzenis and Prof. Reneker (*PCT/US99/11755, 1999*) used electrospun nanofibers in between laminas of a laminate to improve delamination resistance. They arranged PBO nanofibers at the interfaces between plies of the laminate without a substantial reduction for the in-plain properties and an increase in weight and/or ply thickness [4].

4.1.2 Nanofiber-Reinforced Composites in Biomedical Application

In 2004, Fong and his group [5–7] published a series of papers on the usage of electrospun nanofibers–reinforced composites in dental restorative matrix. In their study, electrospun nylon 6 nanofiber nonwoven mats were impregnated with the dental methacrylate of BIS-GMA/TEGDMA (50/50, mass ratio: The molecular structures of dental monomers and initiators can be found in Fig. 4.1) for the restorative composite resins. Three-point bending test demonstrated that flexural strength (FS), elastic modulus (EY), and work of fracture (WOF) of the nanofiber-reinforced composite resins were considerably increased (FS by 36 %, EY by 26 %, and WOF by 42 %) with a small amount of nylon 6 nanofibers sheet (5 wt. %, fiber diameter range 100–600 nm). They also found that the fracture surface of the fiber-reinforced composite resin was very rough, and the fracture surface of the neat resin was relative smoother with much larger fracture steps (Fig. 4.1). They suggested that the presence of nanofibers could effectively deflect the crack. As the crack finally broke away from the nanofibers, numerous fracture lines and steps were created on the fracture surface, indicating the energy consumption during breaking. Most importantly, they realized that the interface bonding between the nylon 6 nanofiber nonwoven and the matrix of BIS-GMA/TEGDMA resin was

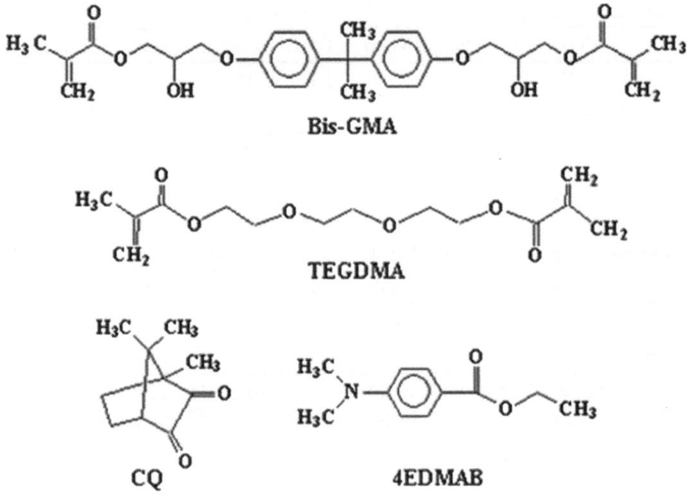

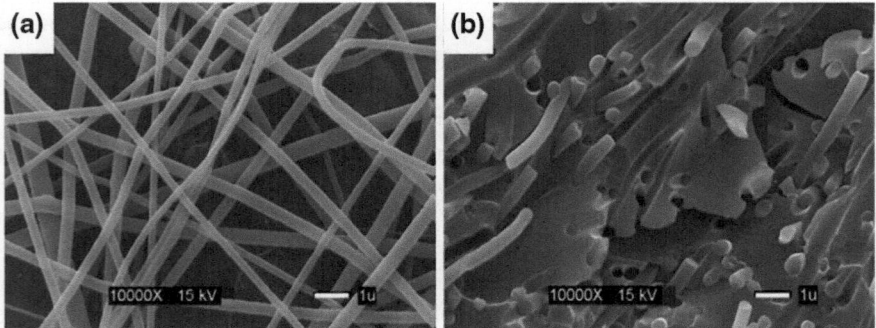

Fig. 4.1 The molecular structures of dental monomers and initiators. Reprinted from Ref. [5], Copyright 2004, with permission from Elsevier

generally poor owing to the few resin remnants could be detected on the end of the pull-out fiber (Fig. 4.2).

To improve the interfacial adhesion between the nanofiber nonwoven and dental methacrylate of BIS-GMA/TEGDMA, Deng and Yang [8] induced the aligned PAN core/PMMA shell (Fig. 4.3) nanofibers into the BIS-GMA/TEGDMA resin owing to the high mechanical resistance of the PAN core and the good adhesion with the matrix (through covalent bonding) of the PMMA shell, thus the composite mechanical properties have been immensely improved (Fig. 4.4).

Fig. 4.2 SEM images of **a** the electrospun nylon 6 nanofiber nonwoven and **b** the presence of the nanofiber in the composite resin, respectively. Reprinted from Ref. [5], Copyright 2004, with permission from Elsevier

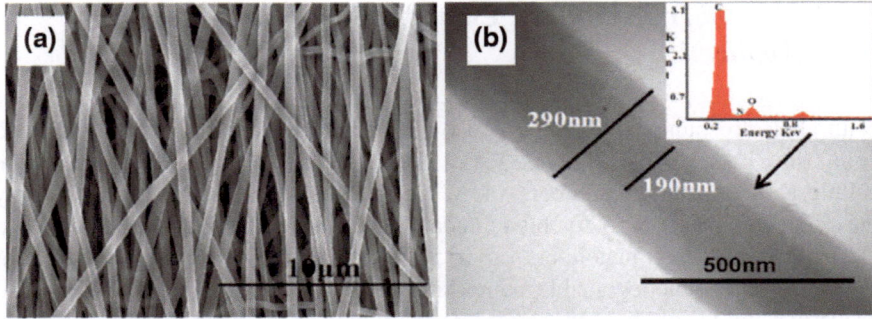

Fig. 4.3 **a** SEM image of the electrospun PAN/PMMA core–shell nanofibers. **b** TEM and EDS result of the PAN-PMMA nanofiber. Reprinted from Ref. [8], Copyright 2010, with permission from Elsevier

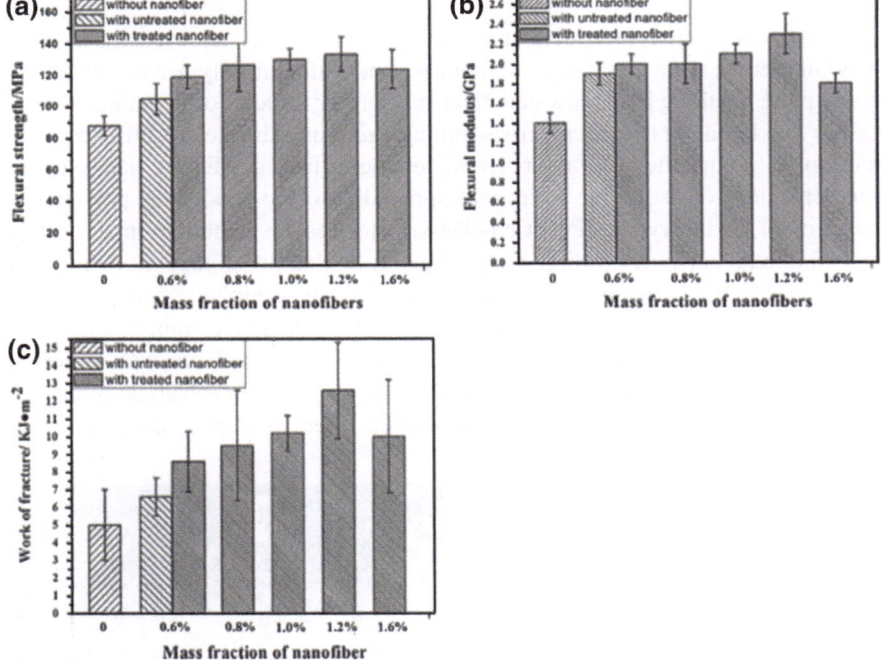

Fig. 4.4 Flexural properties of PAN/PMMA nanofiber/Bis-GMA/TEGDMA composite **a** flexural strength **b** flexural modulus, and **c** work of fracture. Reprinted from Ref. [8], Copyright 2010, with permission from Elsevier

4.1.3 Nanofiber-Reinforced Composites in Biodegradable Application

Natural electrospun nanofibers have also been investigated in nanofiber-reinforced composite. For example, Han and Park [9] explored the potentialities of ultra-fine cellulose electrospun nanofibers (a. v. 560 nm) as reinforcing agent of poly butylene succinate (PBS) for biodegradable composite. Tang and Liu [10] used electrospun cellulose nanofibers (a. v. 500 nm) to reinforce poly vinyl alcohol (PVA) films for biodegradable composite and evaluated composite mechanical properties and visible light transmittance in relation to the fiber content (Fig. 4.5).

4.1.4 How to Improve the Interface Bonding Within the Electrospun-Reinforced Composites

Although many papers on nanofiber reinforcement have been reported, the interface bonding between the electrospun nanofibers and different polymer matrix is generally poor, resulting in the low nanofiber-reinforcing effect. To solve this problem, Özden and Papila [11] induced cross-linking agent into the electrospun nanofiber-reinforced composites. In their study, ethylenediamine (EDA) acted as the supplementary cross-linking agent was sprayed onto the copolymer polystyrene-co-glycidyl methacrylate [P (St-co-GMA)] electrospun nanofiber mat prior to embed into an epoxy resin (Fig. 4.6a). They found that the storage modulus of the P(St-co-GMA)/EDA nanofiber-reinforced epoxy resin exhibited about 10 and 2.5 times higher than that of neat and P(St-co-GMA) nanofiber-reinforced epoxy, respectively, even though the weight fraction of the nanofibers was as low as 2 wt. % (Fig. 4.6b). They suggested that the superior mechanical properties were attributed to the inherently cross-linked fiber structures and the surface

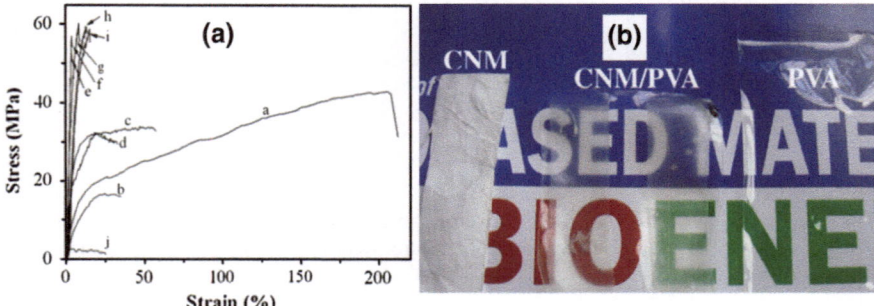

Fig. 4.5 **a** Stress–strain curves of CNM/PVA composite films. Mass content of CNM in the composite films is (a) 0, (b) 4.7, (c) 5.5, (d) 8.4, (e) 23, (f) 34, (g) 40, (h) 50, (i) 60, and (j) 100, respectively. **b** Appearance of film. Mass content of CNM in the composite film is 40 and the light transmittance is 75 %. Reprinted from Ref. [10], Copyright 2008, with permission from Elsevier

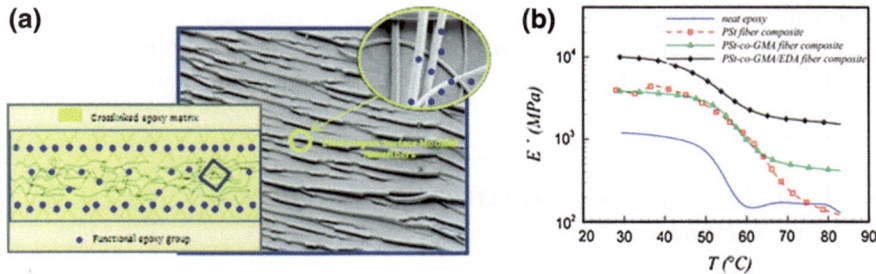

Fig. 4.6 **a** Schematic illustration of the P(St-co-GMA)/EDA nanofiber-reinforced epoxy resin. **b** Storage modulus versus temperature, reinforcement with P (St-co-GMA) with/without amine-sprayed nanofiber, and PSt nanofiber-reinforced composites in contrast to neat epoxy. Reprinted with the permission from Ref. [11]. Copyright 2010 American Chemical Society

modification/chemistry of the electrospun fibers, resulting in strong cross-linked polymer matrix–nanofiber mat interfacial bonding.

4.1.5 Future Challenges

In general, electrospun nanofiber reinforcement offers great opportunities for thin films and bulk samples. However, many important issues on nanocomposites reinforced with electrospun nanofibers have not been solved yet.

1. The modeling and simulation of the mechanical properties of nanofiber composites are still needed to be verified for the lack of mechanical behavior of single-electrospun nanofiber.
2. The control of the nanofiber orientation and the pulverization of nanofibers in the polymer matrix have not been solved yet.
3. The mechanical performances (e.g., strength and toughness) of electrospun nanofibers are still very poor for the fast evaporations and should be improved further.
4. The relationship between the mechanical performances and the fiber diameters is still not very clear.

4.2 Filtration

Filtration is very important in many engineering fields. It was estimated that future filtration market would be up to US \$700b by the year 2020 [12]. For high filtration efficiencies, it is generally necessary that the sizes of the channels and pores in the filter materials should match the scale of the particles or droplets that are to be captured in the filter. Taking the advantage of very high specific surface area, good interconnectivity of porous, low basis weight, high permeability, and controllable pore sizes of the electrospun nanofibers, electrospun nanofiber nonwovens have been widely used in air and liquid filtration applications.

4.2.1 Basic Research on Electrospun Nanofibers in Filtration

Usually, the particle filtration occurs via multiple collection mechanism such as sieving, direct interception, inertial impaction, diffusion, and electrostatic collection. In practice, sieving is not an important mechanism in most air filtration application. Additionally, commercial nanofibers are electrically neutral. So, the remaining important mechanisms in mechanical filtration are direct interception, inertial impaction, and diffusion. The reasonable approximations of filtering media performance have been made using single-fiber filtration theory.

The single-fiber efficiency for direct interception $E_R = (D_P/D_F)^2/Ku$, where D_P is the particle diameter, D_F is the fiber diameter, and Ku is the Kuwabara constant. It can be clearly seen that the filtration efficiency due to the direct interception E_R is inversely proportional to the square of the fiber diameter.

The single-fiber efficiency for inertial impaction $E_I \propto St/(2Ku^2)$.

where St (Stokes number) $= S_D/D_F$ and where S_D is the stopping distance. As can been seen, filtration efficiency due to inertial impaction E_I is inversely proportional to the fiber diameter.

The single-fiber efficiency for diffusion E_D is $E_D = 2.7/(Pe)^{2/3}$.

where Pe is the Peclet number and defined as $Pe = D_F U/D$ and where U is velocity and D is the coefficient of diffusion. Lowering the fiber diameter can increase filtration efficiency for the fast diffusion.

From those dominant filtration mechanisms, it can be found that decreasing fiber diameter in certain filter media can improve the most penetrating particle size (MPPS).

In 2006, Podgórski and co-workers [13] verified those expectations by testing different filters (one microscale filter and five nanoscale fiber filters) as shown in Table 4.4.

The whole experimental setup of filters' testing can be found in Fig. 4.7. In this experiment, desired aerosol flow rate was obtained by two fans, which were placed at the inlet (1) and outlet (13) of the system, respectively. Their rotational speeds were adjusted utilizing the electronic speed controllers (2) and (12). The air flow was determined indirectly on the basis of the pressure drop across a calibrated throttling valve (11) with controllable clearance, which was measured by an electronic manometer (10). The sebacic acid-bis (2-ethylhexyl) ester particles were used as the challenge aerosol generated by a liquid nebulizer with internal air supply pump (4). After generation, the aerosol was diluted by the air purified by passing in through the filter (3). The tested filters were fastened to the wire mesh to give them a

Table 4.4 Calculated filters' performance at air face

Filter no.	MPPS (rm)	P_{max} (dimensionless)	QF_{min} (Pa^{-1})	Δp (Pa)
1	366	0.747	1.69×10^{-3}	173
2	199	0.293	2.85×10^{-3}	431
3	140	0.0222	4.32×10^{-3}	882
4	54	3.28×10^{-4}	6.20×10^{-3}	1295

MPPS: most penetrating particle size
P_{max}: The maximal value of the aerosol penetration
QF_{min}: The calculated minimum value of the quality factor
Δp: Pressure drop. Reprinted from Ref. [13], Copyright 2006, with permission from Elsevier

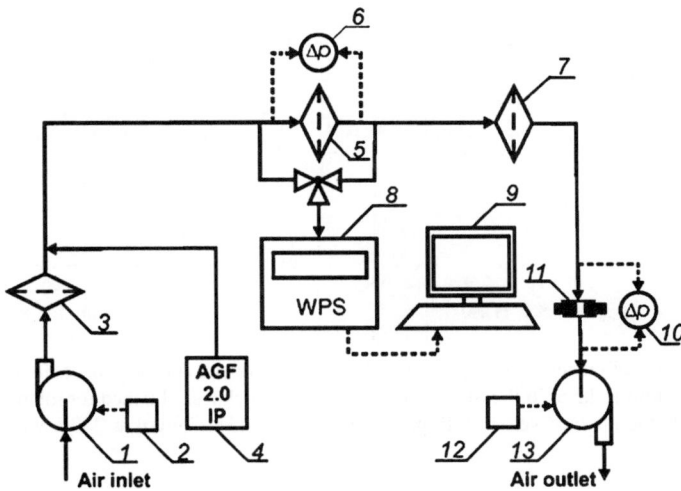

Fig. 4.7 Schematic illustration of the experimental setup for filters' testing. Reprinted from Ref. [13], Copyright 2006, with permission from Elsevier

form of pleated filters, and then, they were placed inside the chamber (5). The pressure drops across these filters were measured by an electronic nanometer (6). The probes for aerosol sampling were located before and after the chamber (5), and with the use of them, the aerosol upstream and downstream of the tested filter were sampled and then fed to the wide-range particle spectrometer (8).

For the poor mechanical strength and durability of nanofibrous media, the authors used a thicker layer of micrometer-sized fibers as substrate to support a thin layer of nanofibers to form a complex filter media, and the macroscopic structural characteristics can be found in Table 4.5.

The efficiencies of the sebacic acid-bis (2-ethylhexyl) ester particles (DEHS) particle removal (size range from 10 to 500 nm) passing through the sets of the backing layer and one of the nanofibrous layers are shown in Fig. 4.8a. It can be found that the using additional layer of the filter composed of nanofibers can

Table 4.5 Macroscopic structural characteristics of the manufactured filters

Type of filter	Thickness L (mm)	Porosity c (dimensionless)	Basis weight pSF (g/m2)
Backing layer (BL)	2.1	0.851	284.9
Nanofibrous layer 1 (NL1)	1.4	0.965	44.4
Nanofibrous layer 2 (NL2)	2.5	0.967	75.1
Nanofibrous layer 3 (NL3)	3.1	0.971	79.4
Nanofibrous layer 4 (NL4)	5.5	0.980	100.5
Nanofibrous layer 5 (NL5)	4.3	0.986	53.0

Reprinted from Ref. [13], Copyright 2006, with permission from Elsevier

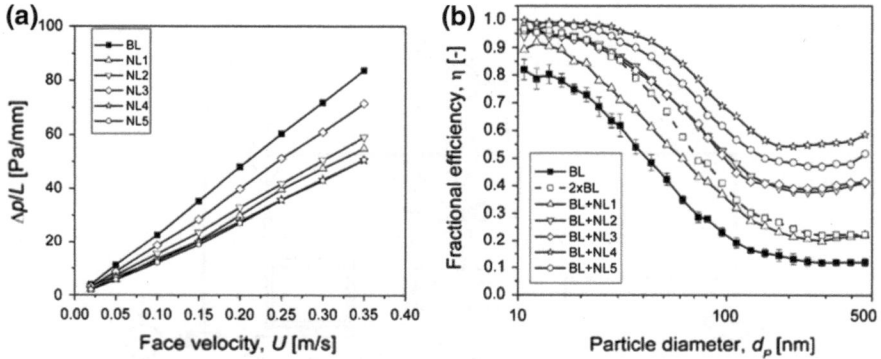

Fig. 4.8 **a** The efficiencies for one and two backing layers and for the sets of the backing layer and one nanofibrous layer. **b** Pressure drop per unit filter thickness obtained for micro/nanofibrous layers versus the air face velocity. Reprinted from Ref. [13], Copyright 2006, with permission from Elsevier

considerably increase the filter efficiency. Additionally, the authors also found that the pressure drop across the filter is linearly dependent on its thickness as shown in Fig. 4.8b.

Till now, many high-efficiency nanofilters made of nanofibers have been realized. For example, Ahn and Lee [14] demonstrated the high-efficiency particulate air (HEPA) filters based on Nylon 6 electrospun nanofibers (diameter: 80–200 nm; basis weight 10.75 g/m^2) exhibited minimum removal efficiency of 99.97 % of particles greater than or equal to 300 nm in diameter. Aussawasathien and co-workers [15] investigated the separation of micro–sub-micron polystyrene particles (0.5, 1, 6, and 10 μm) from water based on electrospun Nylon 6 nanofibers (diameter 30–100 nm). As the diameter of the polystyrene particles is greater than or equal to 1 μm, the removal efficiency can be 100 %. When the diameter of polystyrene is 0.5 μm, the removal efficiency is higher than 90 %. Zhang and Shim [16] systematically investigated the relationship between the electrospun parameters (e.g., concentration of polymer, distance between the tip of syringe and the collector and the flowing rate of the pumper) and filtration efficiency.

4.2.2 Antimicrobial Air Filter

It is well known that heating, ventilating, and air conditioning (HVAC) air filters usually operated in dark, damp, and ambient temperature conditions, which is susceptible for bacterial, mold, and fungal attacks, resulting in unpredictable deterioration and bad odor. To solve this problem, functionalization of the surface of filtering media with antimicrobial agents for long-lasting durable antimicrobial functionality is of current interest. In 2007, Jeong and Youk [17] explored the electrospun polyurethane cationomer (PUCs) nanofiber mats with different amounts of quaternary ammonium groups in antimicrobial air filter. They found that PUCs exhibited very strong antimicrobial activities against *Staphylococcus aureus* and *Escherichia coli* (Fig. 4.9). Ramakrishna and co-workers [18] induced the silver nanoparticles based on different electrospun polymer [cellulose acetate (CA); polyacrylonitrile (PAN); and polyvinylchloride (PVC)] nanofiber for antimicrobial functionality owing to the remarkable antimicrobial ability of silver ions and silver compounds (Fig. 4.10).

4.2.3 Future Challenges

Although electrospun nanofibers have been widely investigated in filtering applications such as antimicrobial air filter, coalescence filter, catalytic filter, highly selective filter, and ion-exchange filter, some filtering performances based on electrospun nanofiber still need to be improved and explored:

Fig. 4.9 Antimicrobial activities of PUCs after incubation for 24 h: (**a**-1) blank and (**a**-2) PUCs against *Staphylococcus aureus,* and (**b**-1) blank and (**b**-2) PUCs against *Escherichia coli.* Reprinted from Ref. [17], Copyright 2007, with permission from Elsevier

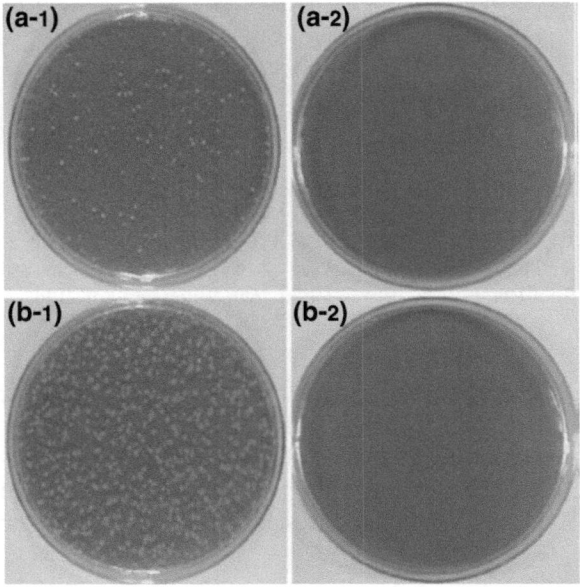

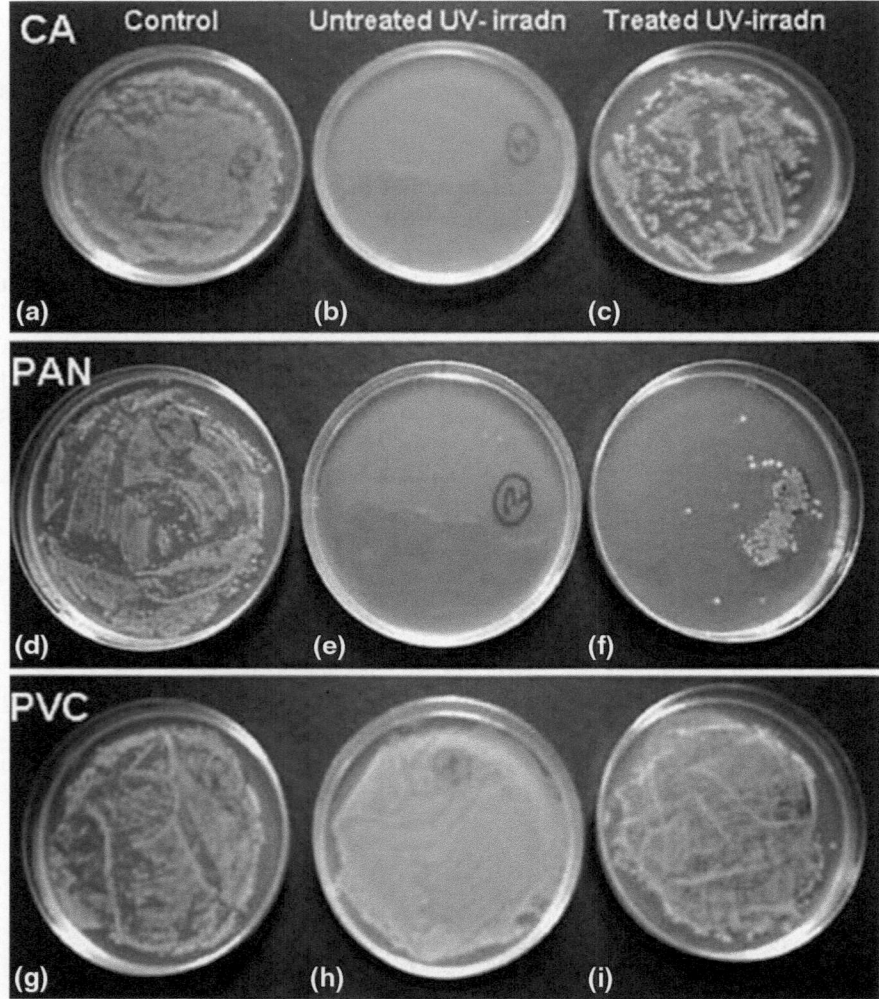

Fig. 4.10 Test result of antibacterial on *E. coli* after 12 h **a** CA as control, incubated with CA nanofibers with 5 wt. % of AgNO₃ before **b** and after **c** UV. **d** PAN as control, incubated with PAN nanofibers with 5 wt. % of AgNO₃ before **e** and after **f** UV. **g** PVC as control, incubated with PVA nanofibers with 5 wt. % of AgNO₃ before **h** and after **i** UV, respectively. Reproduced from Ref. [18] by permission of John Wiley and Sons Ltd

1. The mechanical strength of electrospun nanofiber membranes is still not sufficient to withstand macroscopic impacts during filtration applications, thus supportive fibrous substrate is still needed
2. All the filtering tests were based on electrospun solidified nanofiber membranes, the filtering tested based on other electrospun nanofiber membranes

with complex structures (hierarchical fiber, nanoporous, and honey comb-structured fiber, fiber-web, etc.) have not been explored

3. The uniformity of surface coating and modification of nanofibrous filtering media are still needed improving.

4.3 Catalysis

It is well known that nanostructured materials have opened new possibilities for creating and mastering nanoobjects for novel advanced catalytic materials. In general, catalysis is a molecular phenomenon and the reaction occurs on an active site. A crucial step in catalysis is how to remove and recycle the catalyst after the reaction. The immobilization of catalysts in materials with large surface area advances an interesting solution to this problem. Taking the large surface area and high porosities, electrospun nanofibers, as a novel catalysts or supports for catalysts, have been widely investigated in catalytic field.

4.3.1 Chemical Catalysts

In 2004, Hou and Reneker [19] firstly demonstrated simple in situ electrospinning route for directly depositing Fe nanoparticles with different sizes on carbon nanofibers as shown in Fig. 4.11. In their study, PAN and Fe (acetylacetonate)$_3$

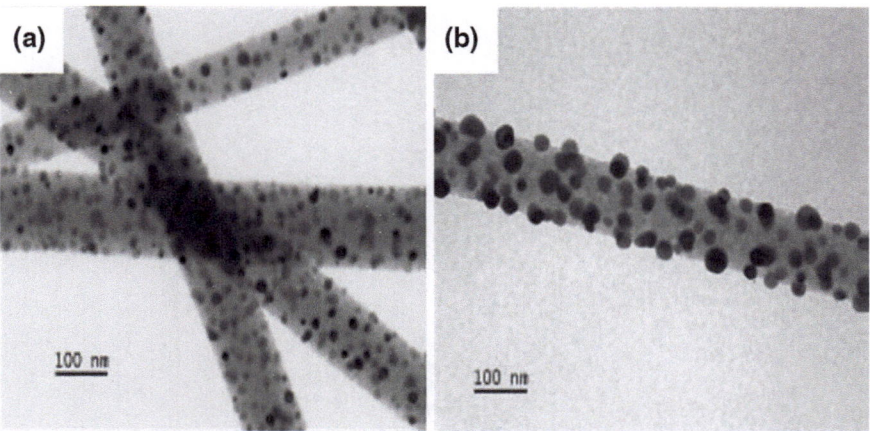

Fig. 4.11 TEM images of the carbon nanofibers containing Fe nanoparticles (*black dots*), made from precursor PAN fibers and Fe (acetylacetonate)$_3$ by calcination. The molar ratio of Fe (acetylacetonate)$_3$ to PAN is **a** 1:2 and **b** 1:1, respectively. The size of Fe nanoparticles is 10 nm **a** and 20 nm **b**, respectively. Reproduced from Ref. [19] by permission of John Wiley and Sons Ltd

were used as the precursors; carbonization and hydrogen reduction were used to convert the precursors into the Fe/carbon composite fibers. With those Fe nanoparticles as chemical catalysts, carbon nanotubes can grow on the surface of carbon nanofibers with controllable length by supplying the hexane vapor with different time as shown in Fig. 4.12.

From 2008, Xia and co-workers [20, 21] published a series of papers on electrospun nanofiber-based supports for chemical catalysis. They used polyol method for decorating diverse metal (Pt, Pd, and Rh) nanostructures (nanoparticles or nanowires) on electrospun TiO_2 (anatase and rutile) and ZrO_2 nanofibers as shown in Fig. 4.13. They also tested the catalytic reaction–based Pt/TiO_2 composite in the hydrogenation of azo bonds in methyl red (Fig. 4.14).

4.3.2 Electrochemical Catalysts

In 2009, Su and Lei [22] investigated the electrochemical catalytical properties based on Pd/polyamide (Pd/PA6) electrospun nanofiber mats for the oxidation of ethanol in alkaline medium in which Pd/PA6 was directly used as electrocatalytic electrodes as shown in Fig. 4.15. High activity and stable performance based on Pd/PA6 have been obtained.

Simultaneously, Kim and co-workers [23] also explored the electrochemical catalytic properties based on electrospun Pt and PtRh nanowires for dehydrogenative oxidation of cyclohexane to benzene. In contrast to the conventional Pt nanoparticle catalysts (e.g., carbon/Pt or Pt black), Pt and PtRh electrospun nanowires electrocatalysts exhibited higher catalytic activities with the same metal loading

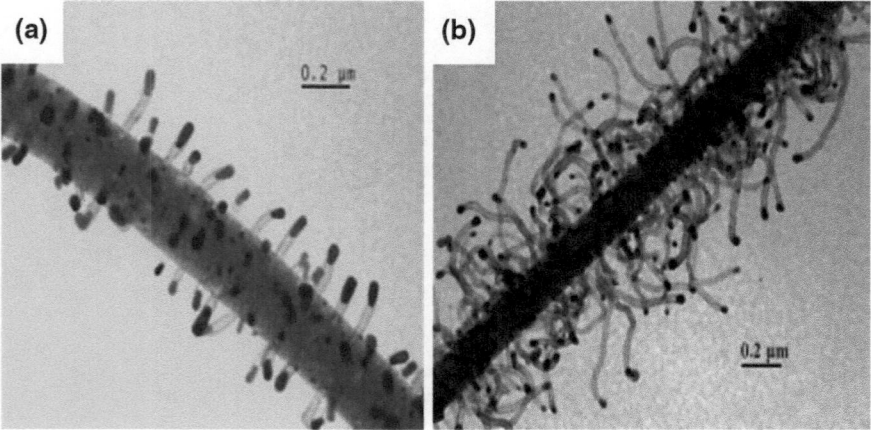

Fig. 4.12 Carbon nanotubes with controllable length on carbon nanofibers by controlling the supplying time of hexane vapor. The time of (**a**) and (**b**) was 3 and 5 min, respectively. The argon flow rate was 600 mL/min. Reproduced from Ref. [19] by permission of John Wiley and Sons Ltd

TiO$_2$ (anatase)

TiO$_2$ (rutile)

ZrO$_2$

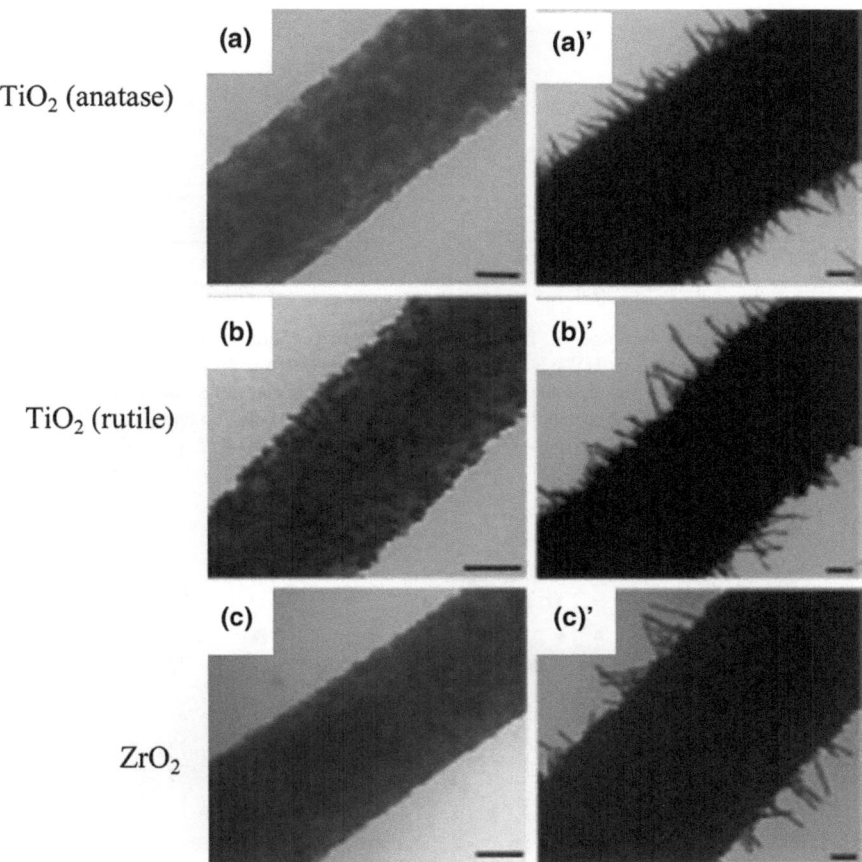

Fig. 4.13 TEM images of ceramic nanofiber decorated with different Pt nanostructures. Reprinted with the permission from Ref. [20]. Copyright 2008 American Chemical Society

amount. Furthermore, PtRh nanowires performed the best catalytic activities with a maximum power density of ca. 23 mWcm^{-2}. Such higher electrocatalytic performances than Pt nanoparticles are attributed to the inherent physicochemical and electrical properties of 1D nanostructures as shown in Fig. 4.16.

1. Nanowire catalysts could provide facile pathways for the electron transfer by reducing the number of interfaces between the electrocatalysts, whereas the nanoparticles are likely to impose more impedance for electrons to transfer particle to particle
2. Adding Rh to Pt to form the PtRh alloy can facilitate the adsorption/desorption properties of benzene and cyclohexane along with the modification of the C–H bond breaking ability
3. The rougher morphology of PtRh nanowires comprise of small nanoparticles, which can provide high catalytic area.

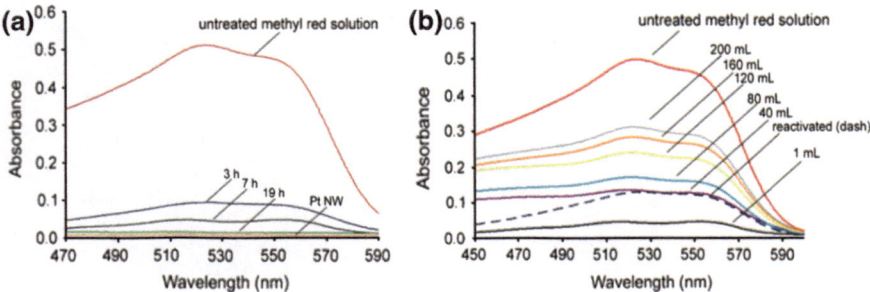

Fig. 4.14 Azo bond cleavage of methyl red via hydrogenation using the decorated fiber membranes. **a** Reduction in UV–vis spectra peak intensity from a 46 μm methyl red solution before and after passing through the anatase/Pt (nanoparticles and nanowires) films. **b** After a gradual increase in the UV–vis peak intensity as more methyl red solution was passed through the anatase/Pt nanoparticles films. Reprinted with the permission from Ref. [20]. Copyright 2008 American Chemical Society

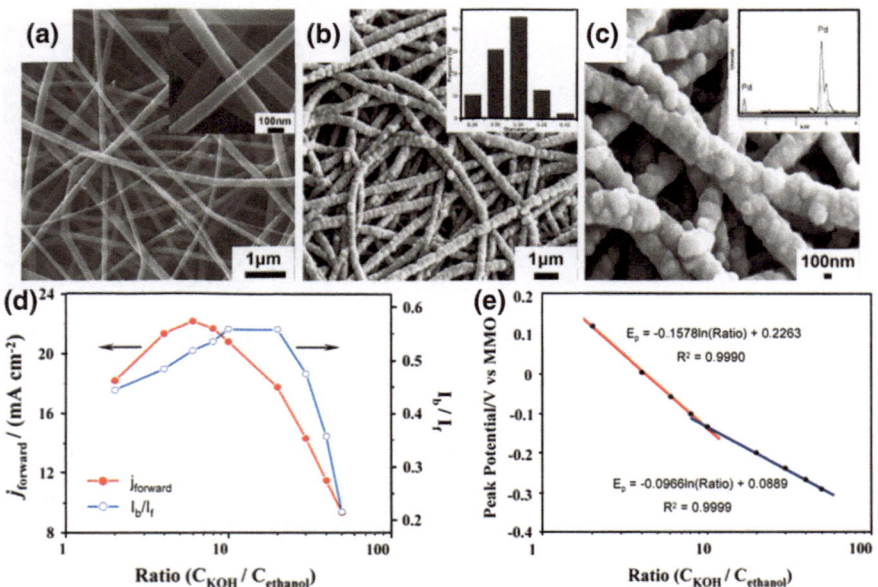

Fig. 4.15 **a** SEM image of PA6 nanofibers. **b** and **c** SEM images of Pd/PA6 nanofibers with low and high magnification. **d** The changes of peak current density and I_b/I_f value with the ratio of KOH concentration to ethanol concentration. **e** Peak potential as a function of $C_{KOH}/C_{ethanol}$. The insets in **a–c** are typical SEM image of PA6 with high magnification, the diameter size distribution of Pd/PA6 nanofibers, and EDX analysis of Pd/PA6 nanofibers, respectively. Reprinted from Ref. [21] by permission of The Royal Society of Chemistry

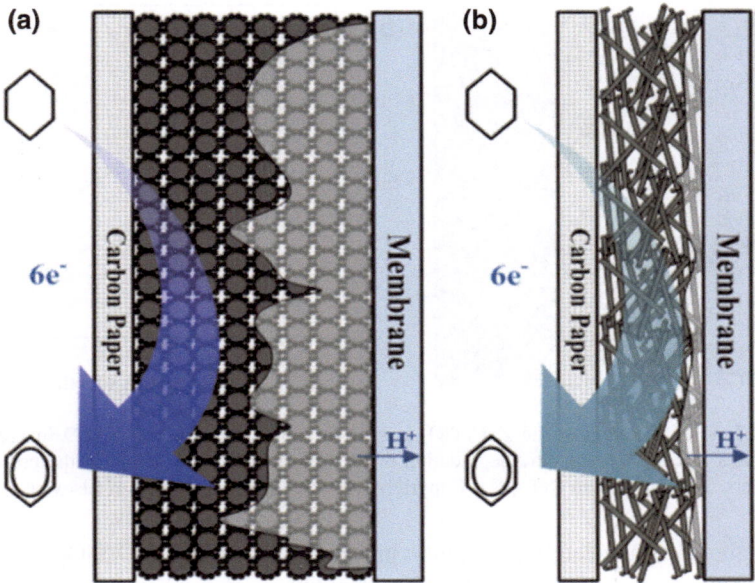

Fig. 4.16 Schematic illustration for cyclohexane electrooxidation over the nanoparticles (**a**) and nanowire (**b**) catalysts. Reprinted from Ref. [23], Copyright 2009, with permission from Elsevier

4.3.3 Photocatalysts

In 2004, Balkus Jr. and co-workers [24] explored the photocatalytic performances based on electrospun titanium dioxide fibers against phenol and formic acid. In their study, pluronic 123 (P123) was used as the template to generate pores within the titanium dioxide fibers. At that time, the photocatalytic performances of the TiO_2 fibers fell well below those of the commercial photocatalysts [Hombikat UV 100 TiO_2 (HK) and Degussa P25 TiO2 (P25)], but their result opened the avenue to use the electrospun materials in photocatalytic field. Following this work, photocatalytic performances based on diverse electrospun ceramic (TiO2, ZnO, SnO2, and so on) nanostructures (nanofibers and nanotubes) have been carried on.

With the development of the photocatalysts, scientists found the fast recombination rate of the photogenerated electron/hole pairs hinders this technique in practice. To solve this problem, three general routes have been developed:

Route 1: Coupling semiconductor of metal oxides with different band-gap widths is an effective route to enhance the photocatalytic activity by increasing the separation of photogenerated electron/hole pairs and extending the energy range of the photoexcitation by virtue of the different Fermi levels of the ingredients in the composite materials. For example, Liu and co-workers [25] explored the photocatalytic performances based on ZnO/TiO_2 composite nanofiber toward the decomposition of Rhodamine B (RhB) as shown in Fig. 4.17. They found that as the ZnO content in the composite was 15.76 wt. %, the composite exhibited the

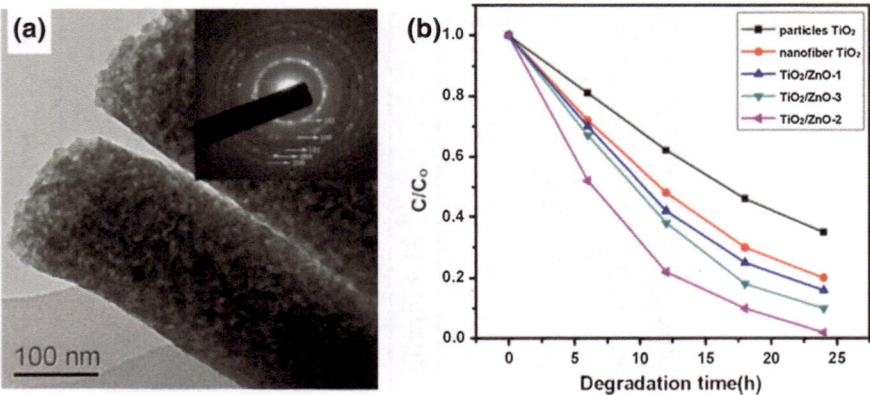

Fig. 4.17 **a** TEM image of the ZnO/TiO₂ composite nanofibers; inset is SAED image of the composites and **b** photocatalytic degradation of RhB in an aqueous solution in the presence of nanofibers. Reprinted from Ref. [25], Copyright 2010, with permission from Elsevier

most efficient photocatalytic activity in contrast to other ZnO/TiO₂ composite fibers, TiO₂ particles, and pristine TiO₂ nanofibers. The enhanced photocatalytic performances can be illustrated as follows: Upon light-activation, the electron transfers from the conduction band of ZnO to that of TiO₂. Simultaneously, the hole transfer from the valence band of TiO₂ to that of ZnO. Thus, efficient charge separation increased the lifetime of the charge carriers and reduces the recombination of the hole-electron pairs in the composite nanofibers, resulting in increasing the quantum efficiency as shown in Fig. 4.18.

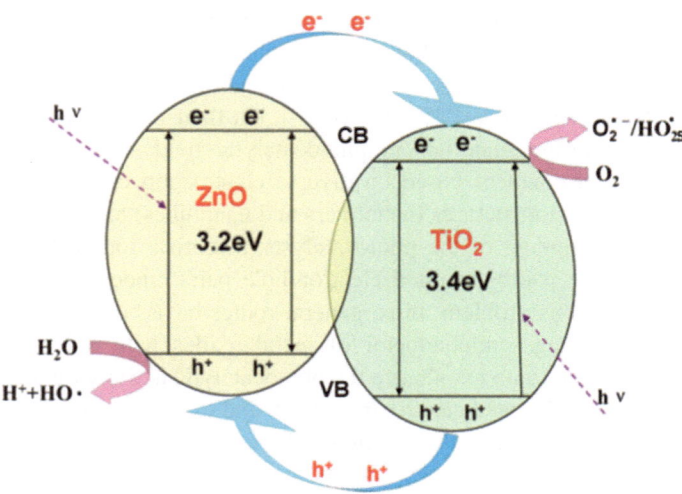

Fig. 4.18 Schematic diagram illustrating the principle of charge separation and photocatalytic activity of the ZnO/TiO₂ composite nanofiber systems. Reprinted from Ref. [25], Copyright 2010, with permission from Elsevier

Route 2: Loading noble metals (Au, Pt, Ag) on the surfaces of semiconductor can also improve the photocatalytic activities, which can reduce the recombination of the photogenerated electron/hole pairs and lengthen the photogenerated electron/hole pairs lifetime through the conduction band electron trapping. For example, Lin and Pan [26] investigated the photocatalytic performances based on Ag/ZnO composite fibers against Rhodamine (RhB). The authors demonstrated that the optimal photocatalytic activity of Ag–ZnO nanofibers could exceed that of pure ZnO nanofiber by a factor of more than 25 (Fig. 4.19a) and the photocatalytic activity of the composite can be tuned by adjusting the Ag contents (Fig. 4.19b). The mechanism can be explained as follows: As the Ag–ZnO systems illuminated by UV light with photon energy higher than the band gap of ZnO, electrons in the valence band can be excited to the conduction band with the same amount of holes left in the valence band. For the work function of ZnO (5.2 eV) is higher than that of Ag (4.26 eV), the electrons will migrate from ZnO to Ag for the Fermi-level equilibration, increasing the lifetime of the photogenerated electron/hole pairs (Fig. 4.19c). Additionally, the Ag nanoparticles can also act as the bridge to transfer the photogenerated electrons to the dye in solution.

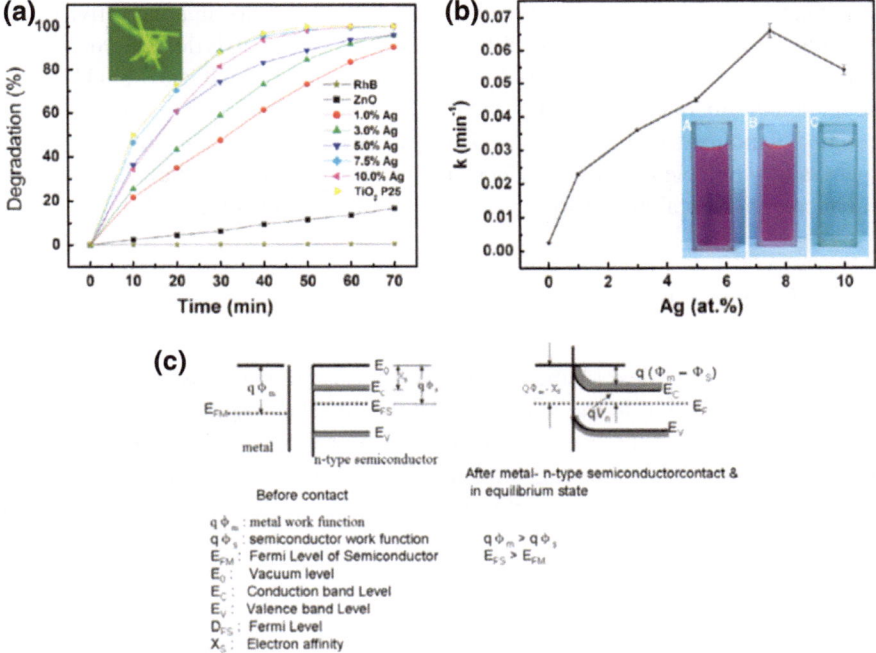

Fig. 4.19 a Kinetics of the photodegradation of an aqueous solution of RhB by Ag–ZnO composite fibers with different Ag contents. **b** Degradation rate constants for composites with different Ag contents. **c** Schematic illustrations of the band structure–related photocatalytic mechanism for Ag/ZnO composite. Reprinted with the permission from Ref. [26]. Copyright 2009 American Chemical Society

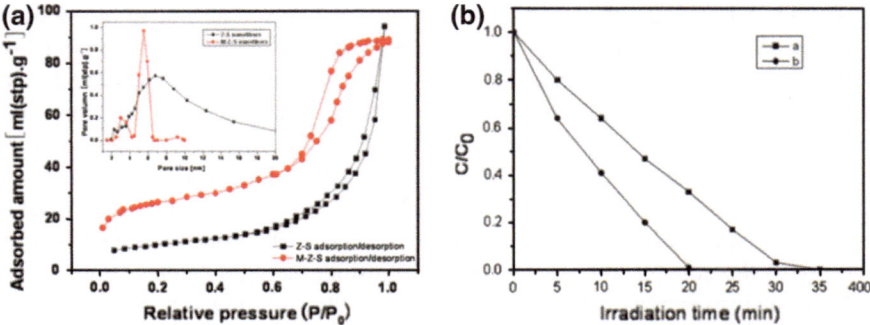

Fig. 4.20 a Nitrogen adsorption–desorption isotherms of Z–S and M–Z–S nanofibers; inset is pore size distribution images of the two nanofibers. **b** The degradation rate of MO over different photocatalysts: (a) Z–S-coupled nanofibers and (b) M–Z–S-coupled nanofibers. Reprinted from Ref. [27], Copyright 2009, with permission from Elsevier

Route 3: Construction of mesoporous structures within the semiconductor to provide large surface area and mesochannels in which the large surface area can offer more active adsorption sites and photocatalytic reaction centers and the mesochannels can act as light-transfer path for introducing incident photo flux onto the inner surface of samples, resulting in enhanced photocatalytic activity. For example, Wang and co-workers [27] compared the photocatalytic performances of both mesoporous $ZnO–SnO_2$ (M–Z–S)-coupled nanofibers and $ZnO–SnO2$ (Z–S)-coupled nanofibers against methyl orange. By adding pluronic P123 into the Z–S-coupled nanofibers (surface area: 57 m^2/g), M–Z–S-coupled nanofibers can be successfully obtained with the surface area of 156 m^2/g (Fig. 4.20a). Additionally, enhanced photocatalytic performances can be also found (Fig. 4.20b).

4.3.4 Future Challenges

Till now, many high-level papers on electrospun nanofibers–based catalysts have been published owing to the good recyclability of electrospun nanofibers, which can effectively avoid the coagulation suffered by nanoparticles. However, in catalytic systems, electrospinning was usually acted as the technique platform, namely, the original catalytic performances of all the electrospun catalysts have been explored before.

4.4 Electronic Devices

It is well known that 1D semiconductor or conductive nanostructures can act as a good model to investigate the dependence of electrical and thermal transport or mechanical properties on dimensionality and size reduction. In addition, they can also be used as interconnects and functional units in fabricating electronic, opto-electronic, electrochemical, and electromechanical devices. Among those 1D nanostructures, electrospun semiconductor or conductive nanomaterials gain special attention for their continuous length and controllable deposited position, which can act as the direct interconnects and functional unit in fabricating diverse devices. Till now, many functional devices have been developed based on electrospun nanomaterials.

4.4.1 Basic Electronic Research on Electrospun Nanostructures

Same to the semiconductor or conductive 1D nanostructures prepared by other methods, 1D electrospun nanostructures with electrical activities have also received much interest in recent years. In 2003, Wang and Santiago-Avilés [28] firstly investigated the electrical performances (zero magnetic field conductivity and magnetoconductivity) based on carbonized carbon nanofibers from PAN nanofibers through a two-probe method. Large negative magnetoresistance (-75 %) was found at 1.9 K and 9 T. Zhou and MacDiarmid [29] electrically characterized the electrospun camphorsulfonic acid (HCSA)-doped polyaniline (PAn)/polyethylene oxide (PEO) composite fibers with the diameters less than 100 nm. They found that as the fibers below 15 nm are electrically insulating through scanning conductance microscopy (SCM) owing to small diameter allow the complete dedoping in air or be smaller than phase-separated grains of PAn and PEO. The I–Vs of asymmetric fibers were rectifying for the formation of Schottky barriers at the nanofiber-metallic working electrode contacts. From then on, a number of electric researching data based on diverse electrospun nanomaterials have been reported.

With the development of electrospinning technique, the electric measurement based on single- or aligned-electrospun nanostructures gains special attention. For example, In 2007, Wang and Santiago-Avilés [30] reported the electric performance based on single–porous electrospun SnO2 ribbon via a two-probe method following a cycle of heating from 300 to 660 K and subsequent cooling from 660 to 300 K. The conductance (G) of the sample is insensitive to the temperature below 380 K and flows an Arrhenius relation with a thermal activation energy of 0.918 ± 0.004 eV from 380 to 660 K; Upon cooling, conductance follows the same Arrhenius relation until 570 K, and another Arrhenius relation with a lower activation energy of 0.259 ± 0.006 eV from 570 to 380 K (Fig. 4.21a). The higher Arrhenius relations are attributed to the surface adsorption and desorption of moisture and oxygen; the lower one is attributed to the particle replacement of adsorbed oxygen by moisture. In 2009, Hou and co-workers [31] compared the electric properties based on aligned carbon nanofibers and found that the aligned carbon nanofibers exhibited anisotropic electrical conductivities (Fig. 4.21b). The differences between the parallel and perpendicular directions to the carbon bundle axes were more than 20 times.

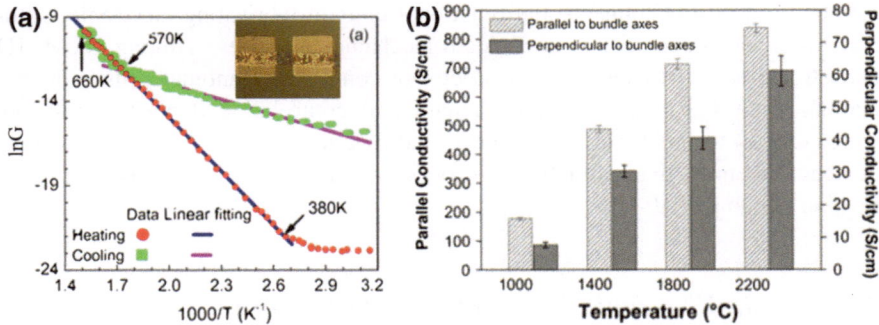

Fig. 4.21 a Arrhenius relation in air and **b** electrical conductivities of aligned-electrospun carbon nanofibers bundles in both parallel and perpendicular directions. **a** Reprinted from Ref. [30], Copyright 2007, with permission from IOP Publishing Ltd; **b** Reprinted from Ref. [31], Copyright 2009, with permission from Elsevier

4.4.2 Field-Effect Transistor

Field-effect transistors (FET), [32–37] sometimes called as unipolar transistor, as the basic building block in logic circuits, have gained special attention, in which an electric field was applied to control the shape and hence the conductivity of a channel of one type of charge carrier in a semiconductor material. The first electrospun-based FET was constructed by Pinto and co-workers [38] with polyaniline/polyethylene nanocomposite fibers as channel as shown in Fig. 4.22.

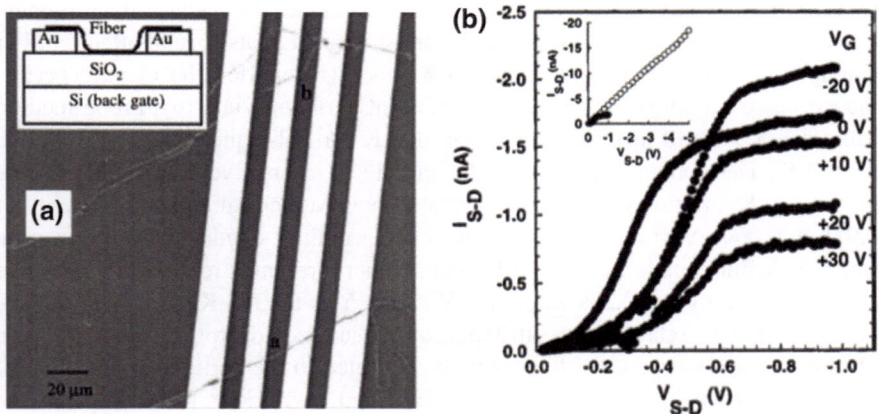

Fig. 4.22 a SEM image of electrospun PANi/PEO fibers over the pre-patterned Si/SiO$_2$ substrate (inset: schematic cross section of the device) and **b** The I–V characteristics of the FET device made up of two nanofibers (lengths 12 and 18 μm, and diameters 300 and 120 nm, respectively.). Reprinted with permission from Ref. [38]. Copyright 2003, American Institute of Physics

Saturation channel currents were observed at low source–drain voltages with the hole mobility of 1.4×10^{-4} cm^2/Vs and the ON/OFF ratio of 2.

From 2006, Pan and his [39] groups published some papers on electrospun inorganic nanofibers-based FET with single-electrospun p-type CuO nanofiber (Fig. 4.23a) and aligned n-type CaN nanofibers (Fig. 4.23b) as channels. They found the conductivity of single p-type CuO nanofiber (a. v. 60 nm) without applied gate voltage was 3×10^{-4} S/cm and single CuO nanofiber–exhibited weak gating effect. Unfortunately, the field-effect mobility of the single CuO nanofiber and aligned CaN had not been calculated.

Recently, inorganic/organic and all-organic electrospun-based FET with high carrier mobilities has been realized by Wang and co-workers [40]. In the case of inorganic/organic system, very high field-effect mobility up to 11.6 cm^{2}V^{-1}s^{-1} could be obtained based on polyacrylonitrile/Au/polyaniline composite fibers without crystallizing the molecular structures of polyaniline (Fig. 4.24). Such high

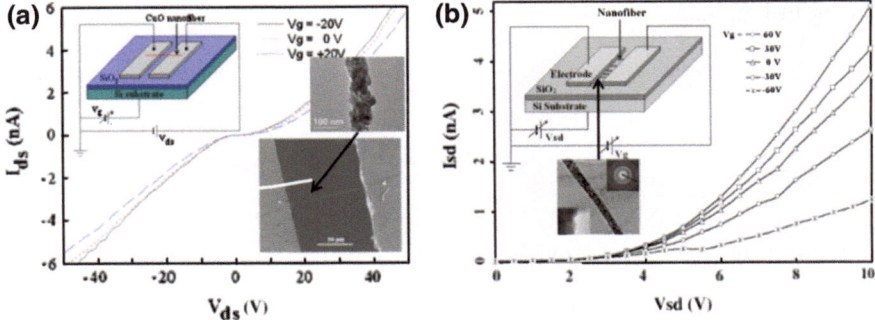

Fig. 4.23 Gate-dependent I–V measurement of single CuO nanofiber (**a**) and aligned CaN nanofibers contacted with Ag electrodes (**b**). Insets are FESEM images of CuO nanofiber (**a**) and CaN nanofiber (**b**), respectively. Reprinted with permission from Ref. [39]. Copyright 2006, American Institute of Physics

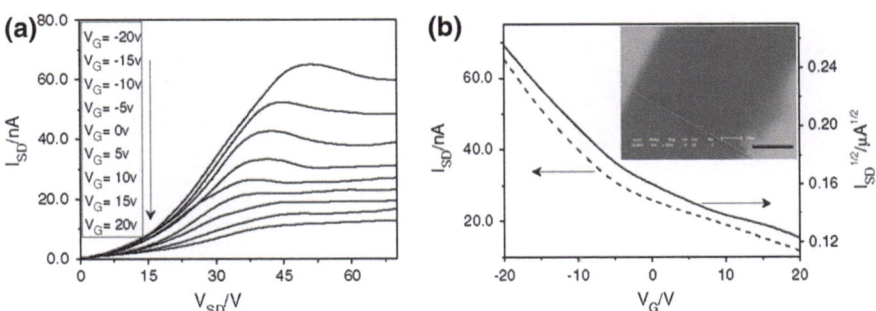

Fig. 4.24 a Output and **b** transfer characteristic of the polyacrylonitrile/Au/polyaniline core–shell nanofiber FET. The inset shows a view of core–shell nanofiber bridging the source–drain gold electrodes. Reproduced from Ref. [40] by permission of John Wiley and Sons Ltd

field-effect mobility is attributed to two effects. One is the 1D nanostructures, which can promote charge transfer and reduce grain boundary effects, resulting in the high field-effect mobility. The other effect is caused by the Au nanoparticles, which can be absorbed electrons within polyaniline, promoting charge transfer between the two parts and making more positive charges left in polyaniline; additionally, Au nanoparticles can act as "conducting bridges" between the polyaniline semiconducting domains, increasing the electrical percolation and promoting the charge transport.

While in the case of all-organic system (Fig. 4.25), polyelectrolyte [sulfonated poly (arylene ether ketone) (SPAEK)]/polyaniline core–shell structure has been used [41]. A high mobility of 3 $cm^2V^{-1}s^{-1}$ is obtained with the current on/off ratio larger than 10^4. Comparing with the previous electrolyte/semiconductor structures, the SPAEK did not act as the gate dielectric but provide internal modulation from the nanointerface of the 1D core/shell structures. Under the negative gate voltage, sulfonic acid on the SPAEK will move into the PANI sheath, resulting in high doping degree by forming $[(PANi) \pm SO_3^-]$ and increasement of drain current as shown in Fig. 4.25c.

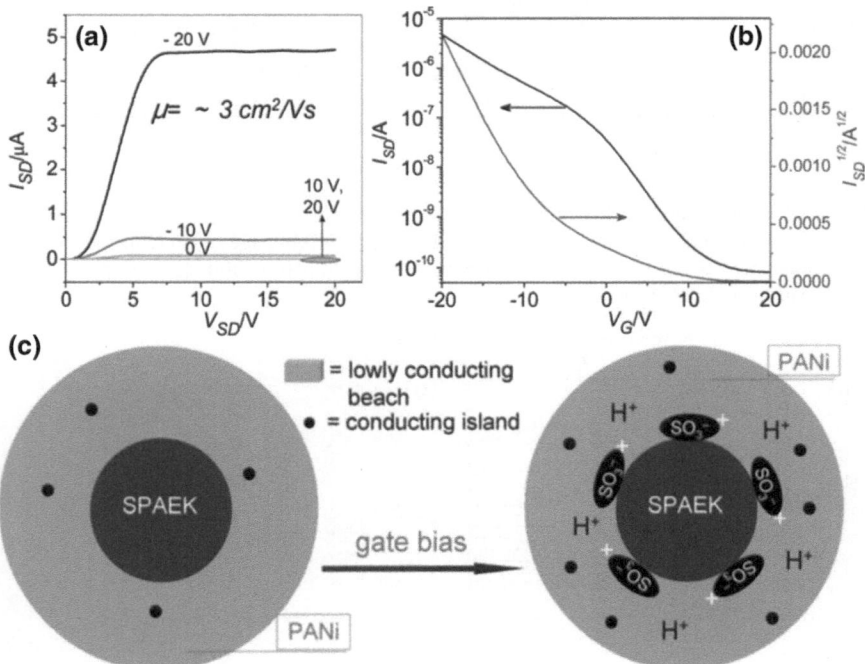

Fig. 4.25 **a** Output and **b** transfer characteristic of the SPAEK/PANi core–shell nanofiber FET. **c** Schematic illustration of the gate-bias-induced charge process in the SPAEK/PANi core–shell nanofibers. Reproduced from Ref. [41] by permission of John Wiley and Sons Ltd

4.4.3 Chemical Sensors

1D nanofibers fabricated by electrospinning have specific surface approximately one to two orders of the magnitude larger than flat films, making them good candidates for sensing applications.

4.4.3.1 Humidity Sensors

Humidity sensors are very important driven by their practical applications in environment monitoring industrial process control, our daily life, and so on. Till known, many humidity sensors based on nanostructure materials have been successfully obtained for the large surface areas of nanoscale structures.

In 2007, Jaruwongrungsee and Lomas [42] firstly explored the electrospun polyaniline micro/nanodots in quartz crystal microbalance humidity sensor, in which electrospun polyaniline micro/nanodots were coated on low-cost quartz crystal microbalance (QCM) sensor and were used to sense the humidity at room temperature. The work principle of such QCM is based on the Sauerbrey equation:

$$\Delta f = - f_0^2 \Delta m / NS\rho \tag{1}$$

where Δf is the frequency change, f_0 is the frequency of the quartz plate, Δm is the weight of applied coating, N is the unique constant of quartz crystal, S is the surface area of coated electrode, and ρ is the density of the quartz plate.

In their study, the humidity control system is a sealed high-quality plastic chamber with low humidity absorption. Three QCM sensors and commercial humidity sensor (SHT-15) are placed in the chamber at the same level; the data of frequency and humidity are sent to the computer via USB port every second. The frequency outputs, humidity, and temperature are recorded in the computer. Water bubbler system is used for the humidity control. Air zero, pure and low-humidity air are split into two lines. The first is passing through mass flow controller A and directly flows into the chamber. The other is passing through mass flow controller B and bubbled through deionized water before entering the chamber. The chamber is filled with air zero flowing at the rate of 1,000 sccm followed by inputting the desired humidity into the chamber by turning on the water-bubbled air zero at the predetermined flow rate and turning off air zero from mass flow controller A. After stabilizing the humidity, the sensing response is recorded. The whole experimental system can be found in Fig. 4.29. The authors found that the as-prepared sensors can respond well to the controlled relative humidity of 40–80 % with short response (3–10 s) and recovery (40–200 s) behaviors. Additionally, the frequency shift of QCM sensor at a certain relative humidity is considerably increased after polyaniline coating and the responses of the sensors are increased with prolongation of the coating time.

A breaking through in the humidity sensors was realized by Li and co-workers [43] with LiCl-doped TiO_2 electrospun nanofibers as sensing elements. In their study, LiCl-doped TiO_2 nanofibers exhibited greatly improved sensitivity with the relative humidity (RH %) range of 11–95 %. In contrast to the previous paper on electrospun humidity sensor, the response and recovery times have been immensely reduced to 3 and 7 s (Fig. 4.26a), respectively. Moreover, good reproducibility (Fig. 4.26b) and stability (Fig. 4.26c) have been obtained. Such excellent sensing performances are attributed to the unique structures of electrospun nanofibers: (1) The large surface of the nanofiber makes the absorption of water molecules on the surface of sample easily; (2) 1D continuous structures of the fibers can facilitate fast mass transfer of the water molecules to and from the interaction region as well as improve the rate for charge carriers to transverse the barriers induced by molecular recognition along the fibers; (3) In contrast to traditional 2D nanoscaled films, the interfacial areas between the active sensing region of the nanofibers and the underly substrate is greatly reduced. Those advantages result in significant gain in the sensing signal and good stability. Inspired by this work, diverse ultra-sensitive humidity nanosensors have been developed based on electrospun nanofibers including pristine metal oxide fibers, basic metal salt-doped metal oxide fibers, coupled metal oxide fibers, and conducting polymer fibers.

Recently, Corres and Matias [44] demonstrated a novel optical fiber humidity sensors-based PVdF electrospun nanowebs in which electrospun PVdF was coated

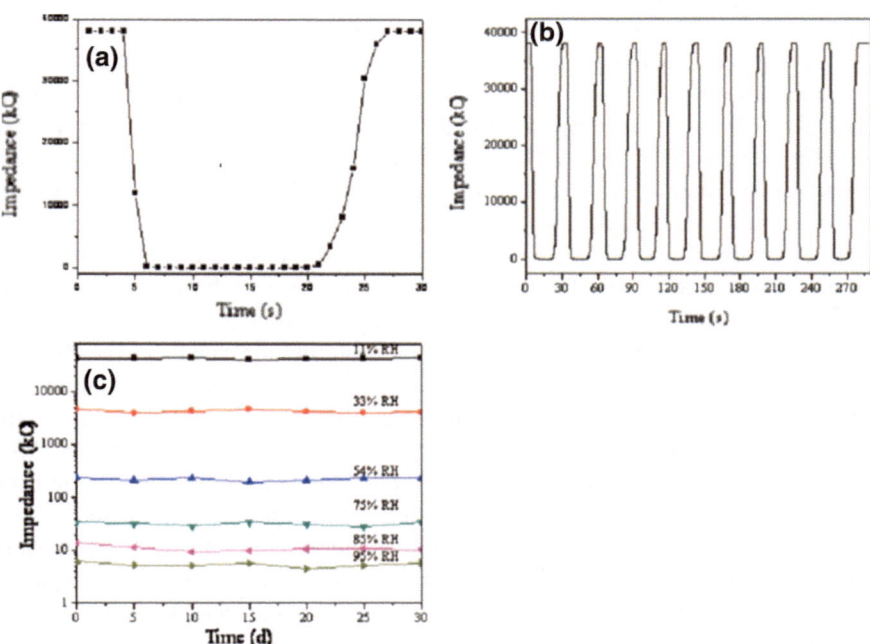

Fig. 4.26 Response and recovery characteristic curves (**a**), reproducibility (**b**), and stability (**c**) of the as-prepared samples. The amount of LiCl within TiO_2 is 30.0 wt. %. Reprinted with the permission from Ref. [43]. Copyright 2008 American Chemical Society

on the surface of a hollow-core fiber (HCF). The as-prepared sensor exhibits a repetitive response in the range from 50 to 70 % relative humidity with a rise time of 100 m. Most importantly, this sensor is intended to be used for monitoring the human breathing. The sensor head was placed at a short distance from the people mouth (5 cm). For the highly porous structure of nanoweb, the rise time of the sensor is around 0.1 s for all sensors with layer thickness in the range 5–7 microns.

4.4.3.2 Gas Sensors

Gas sensor, as a special device to detect the presence of different gases, gained special attention for monitoring environmental pollution. Usually, gas-sensing performance features such as sensitivity, selectivity, response and recovery, stability, durability, reproducibility, and reversibility are influenced by the intrinsic properties of the sensing materials used. Till now, many attempts are carried out to prepare gas sensors based on electrospun nanofibrous membranes including acoustic wave, resistive, photoelectric, and optical gas sensor.

In 2004, Liu and Craighead [45] explored the gas-sensing performances-based single-polymeric polyaniline/poly(ethylene oxide) (PANI/PEO) nanowire. In their study, scanned-tip electrospinning method had been used for depositing isolated and oriented polymeric composite nanowire on lithographically defined micro-electrodes (Fig. 4.27a). Ultra-low detecting limitation (0.5 ppm) against NH_3 can be obtained (Fig. 4.27b). It is very important to note that in their study, the real model on the response times of nanosensor with different diameters corresponding to radius-dependent differences in the diffusion time of NH_3 gas into the electrospun nanofibers has been provided for the first time (Fig. 4.27c). As the diameter of the sample is 335 nm, the response is 74.8 s, which is shorter than that (162.8 s) of sample with the diameter of 490 nm.

Simultaneously, Ding et al. [46] applied electrospun nanofibrous membranes on QCM electrodes as highly sensitive gas sensor with polyacrylic acid (PAA) and poly (vinyl alcohol) (PVA) blends fiber (containing different amount of PAA) as model against NH_3. In their study, sensing experiments were examined by measuring the resonance frequency shifts of QCM which due to the additional mass loading. PVA, in this experiment, acted as the template for forming composite fibers. The average resonance frequency shifts of fibers coated QCM with 11, 18, 25, 33 wt % of PAA to PVA were 40, 150, 240, and 380 Hz, respectively, which is higher than those of casted film-coated QCM sensors (Fig. 4.28a). But all the frequencies of samples could not return to the original position after absorption and desorption processes for the excessive NH_3 were kept inside the 3D structures (Fig. 4.28b).

In addition to those polymeric-based gas sensors, electrospun ceramic-based gas sensors have been also explored by many groups. For example, Sawicka and Gouma [47] explored the sensing performances-based electrospun MoO_3 and WO_3 fibers with NO_2 gas as target. Hao and Peng [48] demonstrated the ethanol-sensing behaviors of the electrospun Cr_2O_3 nanofibers. Although those electrospun

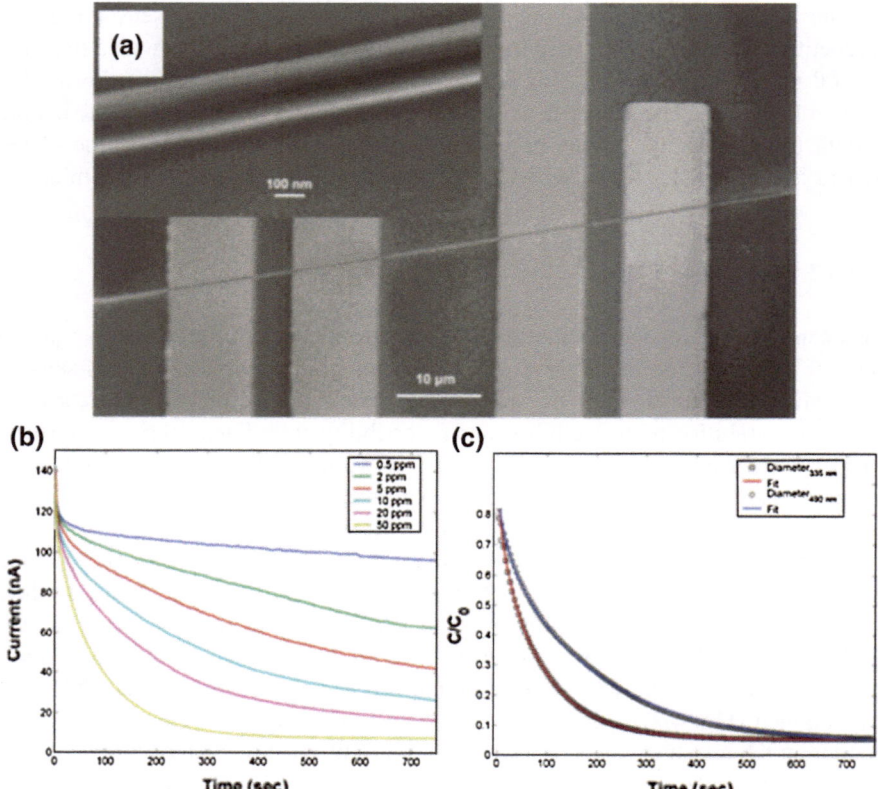

Fig. 4.27 **a** SEM of single PANI nanowire lying across gold electrodes. **b** Response to different concentrations of NH$_3$ gas based on single nanowire (335 nm). **c** Two nanowires with different diameters against NH$_3$ gas. Reprinted with the permission from Ref. [45]. Copyright 2004 American Chemical Society

ceramic-based gas sensors exhibited better sensing performances (e.g., sensitivity, response and recovery behavior) when compared with thin film structures of the same materials, the reported results still do not meet the requirement of ultra-sensitive sensors capable of detecting trace concentrations of gases in the parts per billion range and below. Furthermore, the adhesion of the nanofibers to the substrate for the good electrical contacts between the fibers and the electrodes is still a challenge unsolved. To overcome those problems, Kim and Tuller [49] combined electrospinning and hot press to fabricate ultra-sensitive gas sensors-based electrospun TiO$_2$ nanofibers. Figure 4.29 illustrates the whole process. After the hot press, TiO$_2$ nanofibers, made up of tiny TiO$_2$ nanoparticles (a. v. 10 nm), can be obtained (Fig. 4.29d), resulting in ready gas accessibility to the exceptionally high surface area (138.23 m^2/g). Such gas sensor can exhibit exceptional sensitivity to NO$_2$ with a detection limit below 1 ppb (Fig. 4.30).

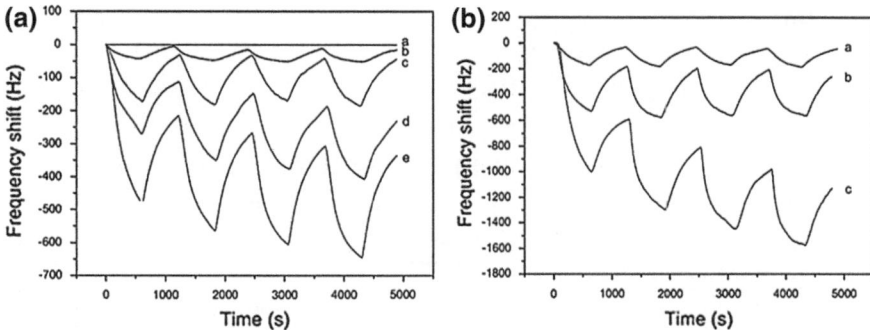

Fig. 4.28 **a** Response of samples with different weight percentage (*a*: 0, *b*: 11, *c*: 18, *d*: 25, *e*: 33 wt. %, respectively) of PAA to PVA exposed to 50 ppm of NH$_3$ at the relative humidity of 55. **b** Response of samples with 18 wt % of PAA to PVA exposed to various concentration (*a*: 50, *b*: 100, *c*: 200 ppm, respectively) of NH$_3$. Reprinted from Ref. [46], Copyright 2004, with permission from Elsevier

4.4.4 Future Challenge

Although many electrospun nanofibers-based electric devices have been successfully obtained, some important issues have not been clearly solved:

1. More experimental studies and theoretical modeling are needed to achieve better control over the size and morphologies of electrospun fibers and a better understanding of relationship between electronic properties and spun fiber morphology.
2. The diversity and scope of materials should be expanded to electrospinning process.
3. The surfaces of all the semiconductor and conducting polymer nanofibers are suspectable to diverse gases and humidity, which makes those nanofibers not stable in fabricating other electric devices (e.g., FET).
4. The diameters of nanofibers within the electrospinning process are not uniform, which makes the electric performances measured based on single-electrospun nanofiber unstable.
5. The low output of the electrospun nanofibers cannot meet the widespread manufacturing demand.

4.5 Lithium-Ion Batteries

Among diverse energy storage devices, rechargeable lithium-ion batteries with high energy density, long cycle lives, flexible design, low self-discharge, high operating voltage, and no "memory effect," are regarded as the effective solution to the ever-increasing demand for high–energy density electrochemical power sources [50, 51].

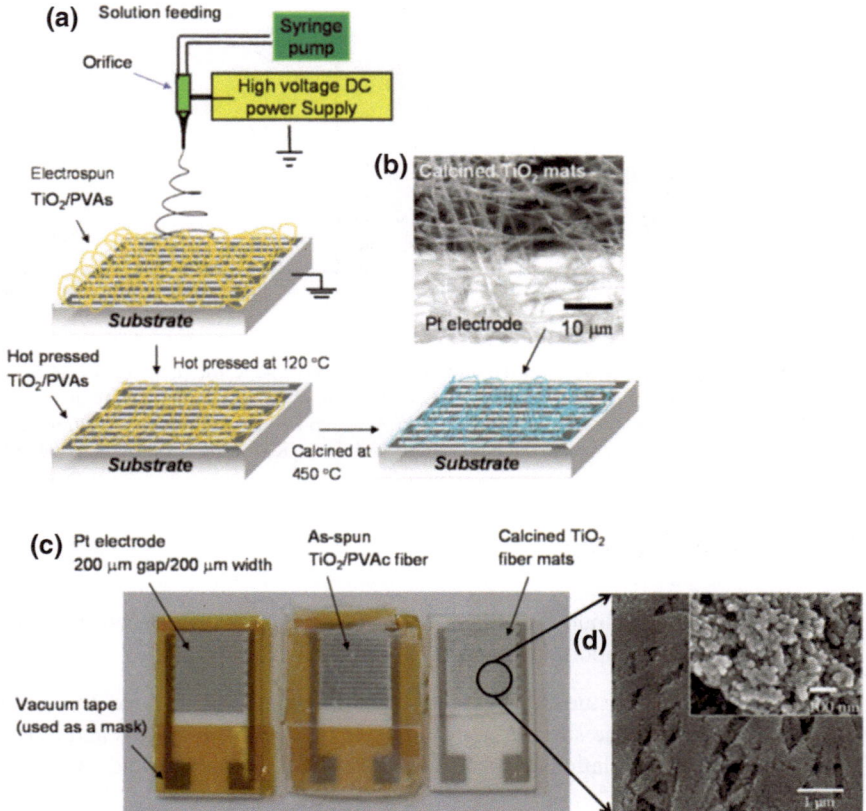

Fig. 4.29 **a** Schematic illustration of the processing steps for depositing TiO₂ nanofibers on Al₂O₃ substrates with interdigitated Pt electrode arrays. **b** Scanning confocal laser micrograph of the calcined samples on top of the Al₂O₃ substrate. **c** Optical micrographs of gas sensor test devices with different processing steps. Reprinted with the permission from Ref. [49]. Copyright 2006, American Chemical Society

Typical lithium-ion battery consists of an anode and a cathode separated by an electrolyte containing dissociated lithium salts, enabling the transfer of lithium ions between the two electrodes as shown in Fig. 4.31. Usually, the electrolyte membrane is a porous separator film to prevent the physical contact between the anode and cathode. As the battery is being charged, external electrical power source injects electrons into the anode. Simultaneously, the cathode gives up some of its lithium ions, which move through the electrolyte to the anode and remain there; thus, electricity is stored in the battery in the form of chemical energy. As the battery is discharging, the lithium ions move back across the electrolyte to the cathode, releasing the electrons to the outer circuit for the external electrical work. In principle, normal powder materials have a long diffusion path for lithium ions and show electrode reaction kinetics, which reduce their practical applications.

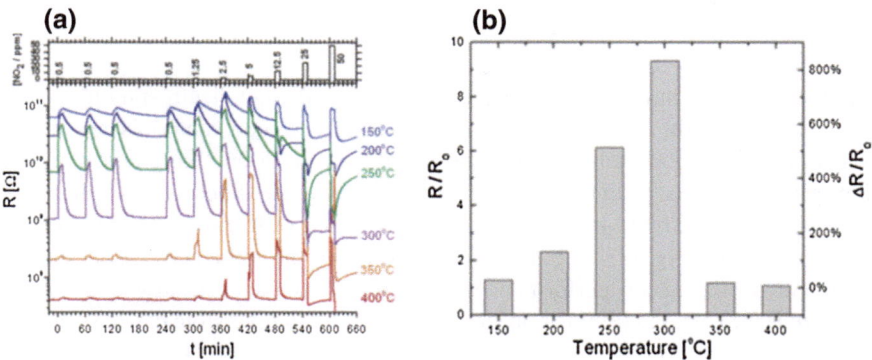

Fig. 4.30 **a** The resistance response during cyclic exposure to 10-min pulses with increasing concentrations of NO_2 mixed in dry air at different operating temperatures. **b** Sensitivity versus temperature upon exposure to 500 ppb NO_2. Reprinted with the permission from Ref. [49]. Copyright 2006, American Chemical Society

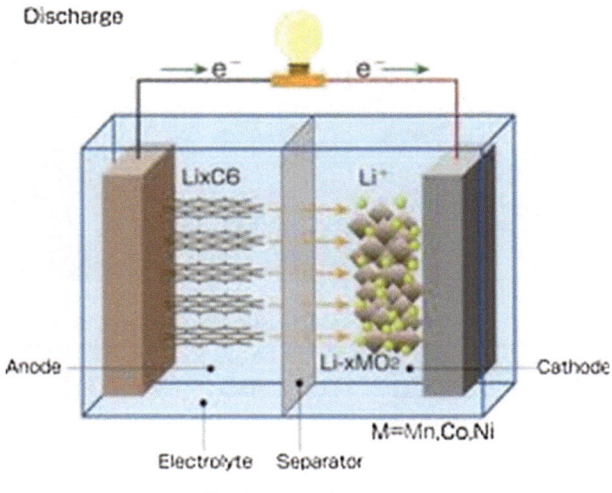

Fig. 4.31 Schematic illustration of lithium-ion battery, consisting of anode and cathode separated by one electrolyte containing dissociated lithium salts. Reprinted from Ref. [50] by permission of Taylor and Francis Ltd

Electrospun nanofibers gain special attentions in fabricating electrodes within lithium-ion battery driven by the shorter diffusion path in contrast to the commonly employed powder materials, fast intercalation kinetics for the large surface-to-volume ratio, and decreased charge-transfer resistance at the interface between the electrolyte and active electrode materials for the large number of lithium insertion sites can be provide based on it. Furthermore, electrospun-based separator within lithium-ion battery also attracts immensely attentions for their small pore size and large porosity, which can decrease the cell resistance.

4.5.1 Electrospun Nanofiber-Based Anodes

Significant efforts have been made to develop high efficient electrospun nanofiber-based anodes within lithium-ion batteries. Usually, anode materials are mostly based on electrospun carbon composite nanofibers. In 2006, Yang and Kim [52] demonstrated the usage of electrospun-derived carbon nanofiber webs for the anode materials within lithium-ion batteries. In their study, different thermal treatments were carried on (Fig. 4.32a, b, and c) to investigate the influences of the thermal treatment on the anode performances (Fig. 30D). As the thermally treated temperature is 700 °C, the lithium-ion storage capacity and reversible capacity are 510 and 300 mAh g^{-1}, respectively. As the temperature is 1000 °C, the lithium-ion storage capacity and reversible capacity are 1000 and 450 mAh g^{-1}, respectively. Such large irreversible capacities are related to the electrolyte decomposition and formation of a solid electrolyte interphase at ca. 0.8 V. After a second cycle, relative stable capacity can be obtained. As the temperature is 2,800 °C, reproducible discharge/charge plateau at about 0.2 V (stage formation) and reversible capacity of

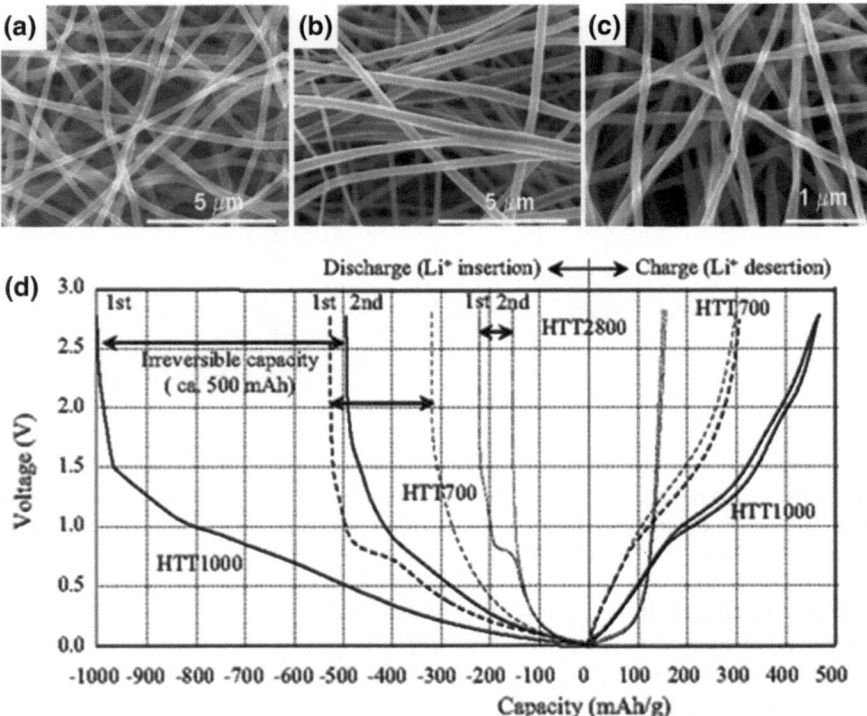

Fig. 4.32 FESEM image of thermal-treated carbon nanofibers **a** 700, **b** 1000, and **c** 2800 °C, respectively. **d** Charge–discharge curves for nanofiber webs thermally treated at 700, 1000, and 2800 °C (current density = 30 mAg^{-1}, second cycle). Reproduced from Ref. [52] by permission of John Wiley and Sons Ltd

130 mAh g^{-1} can be observed. *It is important to point out that the carbon nanofiber treated at* 1,000 °C *exhibits a reversible capacity of* 450 mAh g^{-1}, *which is larger than both the* 372 mAh g^{-1}, *theoretical value for graphite.*

With the development of electrospun carbon-based anode in lithium battery, scientists found a very controversial phenomenon: Li-ion storage capacity of carbon-based anode tended to decrease with increasing the heating-treatment temperature due to the reduction in hydrogen content and the reduction in specific surface area from pores; however, lower the heating-treatment temperature, a rather high resistance and a large irreversible capacity will emerge. To overcome this problem, novel strategy has been developed to retain the high capacity of the low-temperature carbon and improve the electrical conductivity. Till now, a series of carbon/composite nanofibers-based anode for lithium-ion battery have been successfully obtained. For examples, Wang and Chen [53, 54] fabricated carbon/cobalt and carbon/Fe$_3$O$_4$ composite nanofibers-based anode for lithium-ion battery. Such composite nanofibers exhibit excellent electrochemical properties. In the case of carbon/cobalt, higher reversible capacity (>750 mAh g^{-1}) even after 50 cycles and better rate capability (578 mAh g^{-1} at 1 C rate) in contrast to pure electrospun carbon nanofibers-based anode can be obtained (Fig. 4.33a and b). While in the case of carbon/Fe$_3$O$_4$, high reversible capacity of 1,007 mAh g^{-1} at 80th cycle and good rate capability (Fig. 4.33c and d) can be obtained. Till now, diverse electrospun carbon composite fiber-based anode in lithium-ion batteries has been obtained including carbon/Si system, carbon/metal (Ni, Cu, Sn), and carbon/metal oxide (MnOx, LiFePO$_4$, Si@SiOx) composite fibers [55–64].

In addition to those electrospun carbon composite fibers-based anodes, electrospun metal oxide fiber-based anodes have also been investigated in lithium-ion batteries. Fan and Whittingham [55] fabricated electrospun manganese oxide nanofiber-based anodes for lithium-ion battery. The capacity on reaction with lithium exceeded 450 mAh g^{-1} for 50 cycles at 0.5 mA/cm^2 can be obtained.

4.5.2 Electrospun Nanofiber-Based Cathodes

Similar to anodes, electrospun nanofibers have been also applied in fabricating cathodes for lithium-ion batteries. In 2005, Chen and co-workers [65] demonstrated electrospun LiCoO$_2$ fiber-based anode. The cyclic voltammogram curves indicated faster diffusion and migration of Li ions. In the first charge–discharge process, initial charge and discharge capacities are 216 and 182 mAh g^{-1}, respectively. After the twentieth cycle, the discharge capacity decreased to 123 mAh g^{-1} (Fig. 4.34a and b) for the dissolution of cobalt and lithium cations escaping form LiCoO$_2$ to form the crystalline LiCO$_2$ and CoF$_2$ impurities. To further improve the stability of LiCoO$_2$ nanofiber-based cathode, Chen and groups [66] fabricated electrospun LiCoO$_2$–MgO core–shell fibers-based cathodes. The cyclic voltammogram experiments indicated that the core–shell fibers

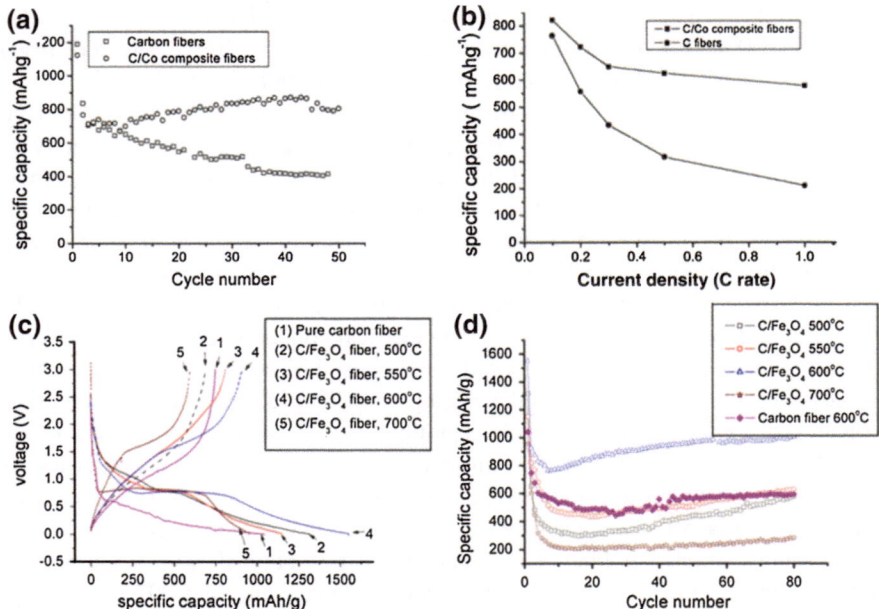

Fig. 4.33 **a** Electrochemical behaviors of electrospun carbon nanofiber-based anode and carbon/cobalt composite fiber-based anodes with the current density of 100 mA g^{-1}. **b** Discharge capacities of the fibers-based anode measured at different current densities from 100 to 1000 mA g^{-1}. **c** Voltage profiles of the first-cycle carbon nanofiber-based anode and carbon/Fe$_3$O$_4$-based anode. **d** Discharge capacity versus cycles of the samples. Reprinted from Ref. [53, 54], Copyright 2008, with permission from Elsevier

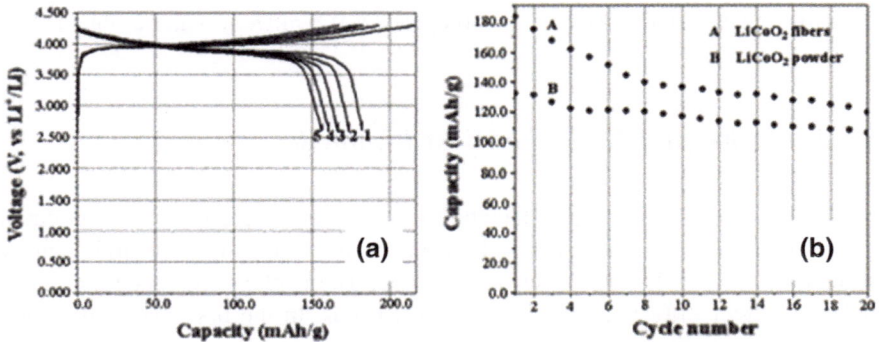

Fig. 4.34 **a** First five charge–discharge curves and **b** cycle life of the LiCoO$_2$ fibers at the current density of 20 mAh g^{-1}. Reprinted with the permission from Ref. [65]. Copyright 2005, American Chemical Society

exhibited improved reversibility, smaller impedance growth, and preferable electrochemical properties in contrast to the bare LiCoO$_2$ fiber-based cathode as shown in Fig. 4.35.

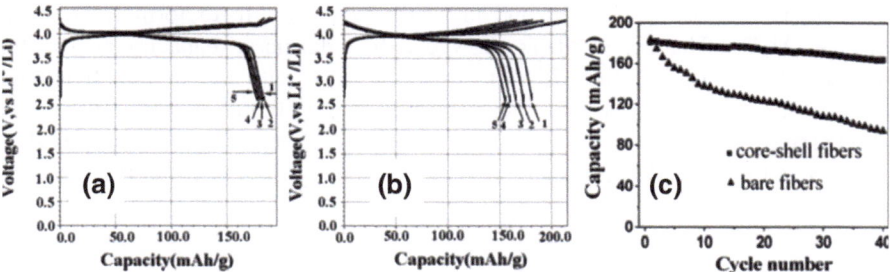

Fig. 4.35 First 5 charge–discharge curves for the core–shell fiber electrode (**a**) and the bare LiCoO$_2$ fiber electrode (**b**), and their cycle performance (**c**). Reproduced from Ref. [66] by permission of The Royal Society of Chemistry

Usually, metal oxide semiconductors have poor conductivity, which leads to high impedance and low rate capability. To overcome this problem, carbon/metal oxide composite fiber-based cathode has been developed in which carbon was used to improve the conductivity of the whole materials. In 2011, Zhang and co-workers [67] demonstrated the electrospun LiFePO$_4$/carbon composite fiber-based cathode from PAN/LiFePO$_4$ precursors. In their study, the amount of PAN (4 wt. %) is fixed and that of LiFePO$_4$ precursors ranges from 4 (S1) to 8 (S2) and to 12 (S3) wt %. Initial charge–discharge curves indicate that flat voltage plateaus can be observed between 3.31–3.54 V and 3.26–3.55 V for the main characteristic of the two-phase reaction of LiFePO$_4$. Additionally, increasement of the carbon amount in the final product can generate higher polarization and less flat plateaus between 2.96 and 3.87 V (Fig. 4.36a). Furthermore, all the LiFePO$_4$/carbon composite fiber-based cathodes show good cyclability (Fig. 4.36b).

Recently, scientists found that partial substitution is an effective method in improving the electrochemical performances of existed cathode materials. Ding and Zhang [68] demonstrated electrospun Al-doped LiNi$_{1/3}$Co$_{1/3}$Mn$_{1/3}$O$_2$ nano-fiber-based cathode. XRD indicated that the Al element was incorporated into the

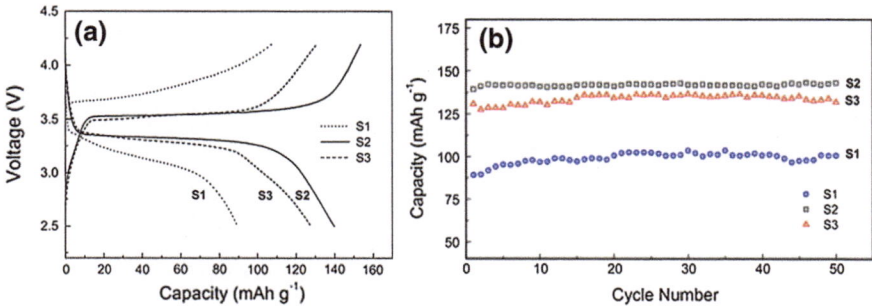

Fig. 4.36 a Initial charge–discharge curves of LiFePO$_4$/carbon composite fiber-based cathode and **b** Cycling performance curves of the three samples. The charge–discharge rate is 0.1 C. Reprinted from Ref. [67], Copyright 2011, with permission from Elsevier

crystal of $LiNi_{1/3}Co_{1/3}Mn_{1/3}O_2$ and takes the place of Mn to form $LiNi_{1/3}Co_{1/3}Mn_{1/3-x}Al_xO_2$. The doped fibers showed a first discharge capacity of 166.32, 172.80, 180.14, 186.59, and 175.78 mAh g^{-1} at 0.1 C with x = 0, 0.02, 0.04, and 0.08, respectively (Fig. 4.37a). Furthermore, as x = 0.06, sample delivered a discharge capacity of 178.03 mAh g^{-1} after 30 cycles (Fig. 4.37b). Simultaneously, all the samples have good reversible capacity even at 2.0 C (Fig. 4.37c).

4.5.3 Electrospun Nanofiber-Based Separators

Separator, within the lithium-ion battery, is a porous membrane between the anode and cathode to prevent the physical contact of the electrodes and enable ionic transport. Traditional commercial separators were mostly based on polymeric (polyolefin, polyamide, nylon, and so on) microporous membranes for their suitable properties such as suitable thickness, mechanical strength, and good chemical stability. But they also suffered low thermal stability, low porosity (around 40 %), and poor wettability with polar liquid electrolyte, resulting in high cell resistance, reduced energy density, and low rate capability within the lithium-ion batteries.

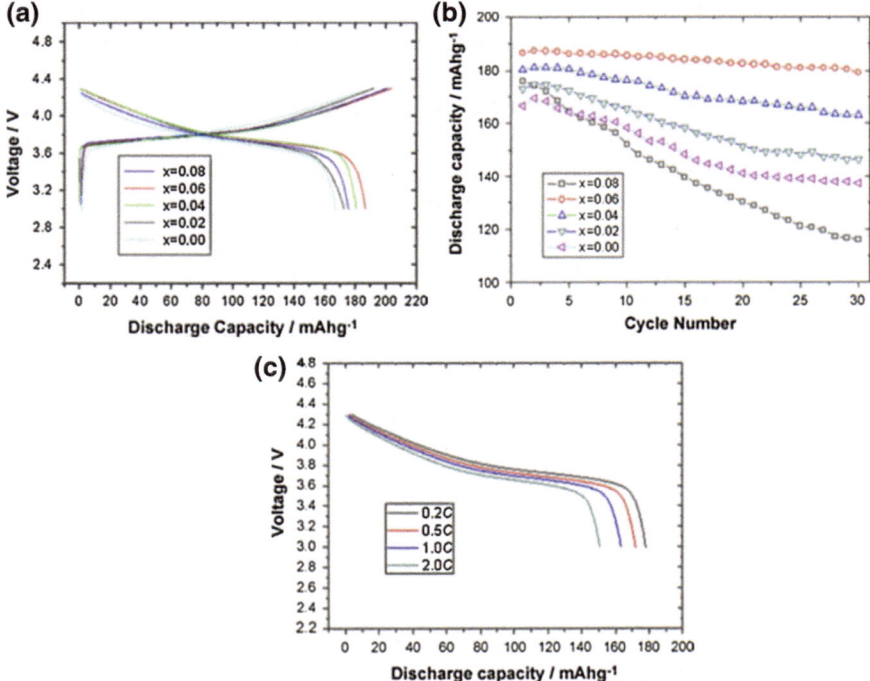

Fig. 4.37 **a** Initial charge–discharge curves of $LiNi_{1/3}Co_{1/3}Mn_{1/3-x}Al_xO_2$ nanofibers, **b** Capacity retention of $LiNi_{1/3}Co_{1/3}Mn_{1/3-x}Al_xO_2$ nanofibers, and **c** Rate capability of $LiNi_{1/3}Co_{1/3}Mn_{1/3-0.06}Al_{0.06}O_2$ nanofibers. Reprinted from Ref. [68], Copyright 2009, with permission from Elsevier

Recently, electrospun nanofiber-based separators have attracted immensely attention for their tunable pore sizes and large porosity, which can provide desirable pathways for high-rate charge/discharge of lithium-ion batteries. In 2006, Gao and co-worker [69] fabricated electrospun PVDF-based membranes (EPMs) for battery separators applications. To improve the physical properties of EPMs, thermal treatment (160 °C for 2 h in a convection oven) has been carried on to treat the EPMs to form an interconnected web structure (Fig. 4.38a). To test the efficacy of the EPMs, the coin cells with the configuration $LiMn_2O_4$/EPM-liquid electrolyte/Li metal are fabricated. Compared with Celgard^{TM} 2400 (commercial PP separator), the cell with EPM shows better cycling ability of CV and charge-discharge performances with little capacity loss even after 50 cycles at $C/2$ rate (Fig. 4.38b, c, and d). In addition to the elelctrospun PVDF-based membranes for battery separators, PAN and P(VDF–HFP) were also applied in separators [70–72].

To further improve liquid electrolyte uptake, electrochemical stability window, and lower the interfacial resistance with lithium on the electrospun polymeric-based separators, Zhang and co-workers [73] incorporated lithium lanthanum titanate oxide (LLTO) into polymeric nanofibers. Compared with Celgard^{TM} 2400 (commercial PP separator) and electrospun PAN-based separators, LLTO/PAN

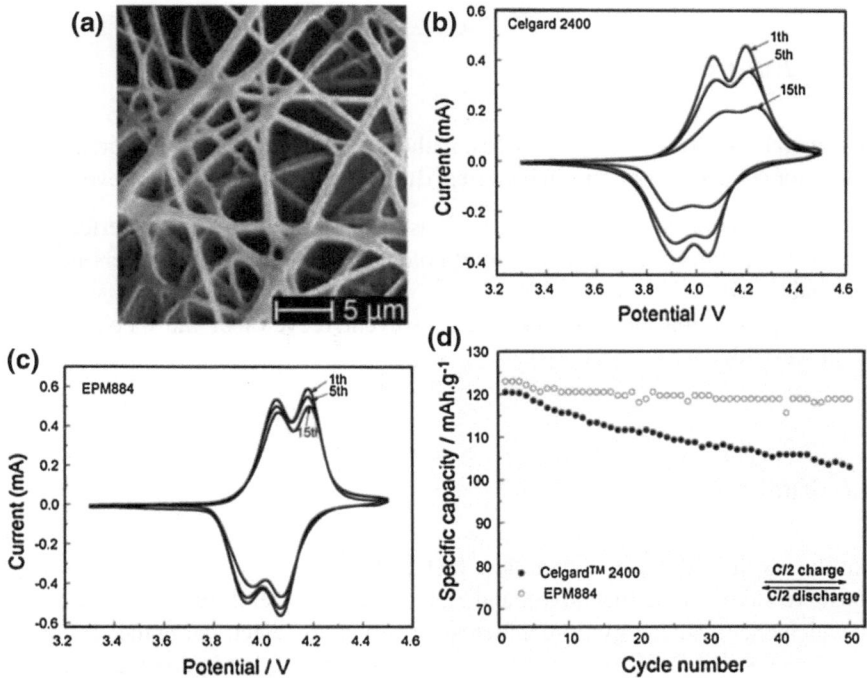

Fig. 4.38 a EPM884 after thermal treatment. **b** and **c** are CV for the cells with Celgard^{TM} 2400 and EPM884 at sweep rate of 0.1 mVs^{-1}. **d** Cycle performance of the cells with Celgard^{TM} 2400 and EPM884 at room temperature. (*EPM884 means the average diameter of the EPM is 884 nm) Reprinted from Ref. [69], Copyright 2006, with permission from Elsevier

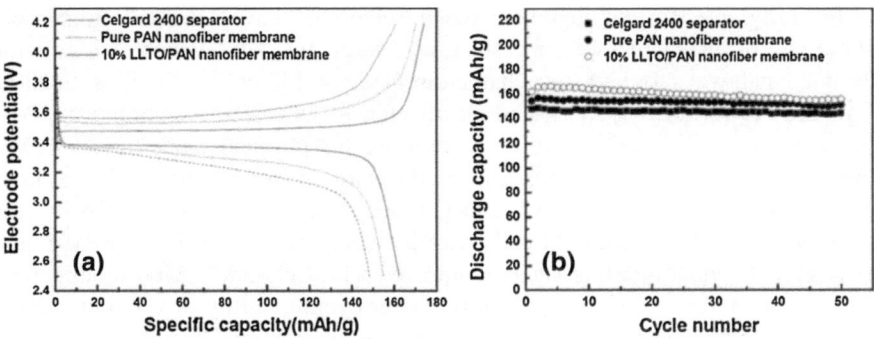

Fig. 4.39 **a** First-cycle charge and discharge curves and **b** Cycle performance of Li/LiFePO$_4$ cells containing liquid electrolyte-soaked Celgard 2400 separator, pure PAN fiber-based membrane and LLTO/PAN composite fiber-based membrane at 0.2 C rate. Reprinted from Ref. [73], Copyright 2011, with permission from Elsevier

exhibited the highest charge/discharge capacity and with the discharge capacity of 156 mAh g^{-1} (91.7 % of the theoretical capacity of LiFEPO$_4$: 170 mAh g^{-1}) subjected to 50 cycles at room temperature (Fig. 4.39).

4.5.4 Future Challenge

In general, electrospun-based anodes, cathodes, and separators offer great opportunity for advanced lithium-ion batteries. But several issues hinder its development:

1. Little theoretical study on the relationship between the electric performances and the nature (e.g., diameter, morphology, structures) of nanofibers has been established.
2. Low production and high cost (e.g., solvent recycle) of the electrospinning process hinders it in industry.

4.6 Fuel Cells

Similar to the lithium-ion batteries, fuel cells also contain three parts: cathode (positive side), anode (negative), and electrolyte. In general, fuel cells are devices in which chemical energy from a fuel is converted into electrical in the presence of catalyst (Fig. 4.40). Hydrogen is the most common fuel, but hydrocarbons such as natural gas and alcohols like methanol are sometimes used. Fuel cells are different from batteries in that they require a constant source of fuel and oxygen to run, but they can produce electricity continually for as long as these inputs are supplied. Fuel cells are mainly classified by the type of electrolyte they used (e.g., proton

Fig. 4.40 Schematic illustration of fuel cells, consisting of anode and cathode separated by electrolyte

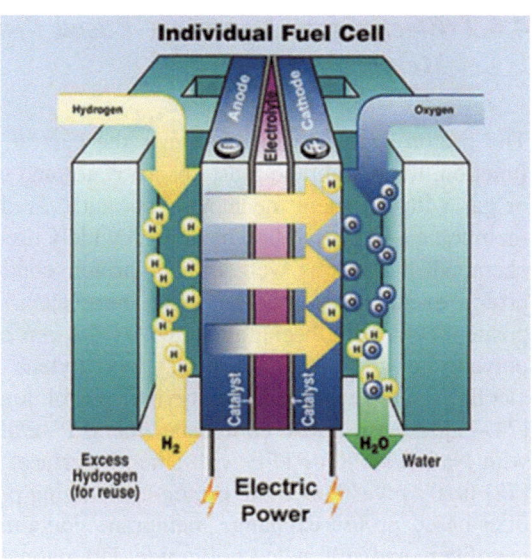

exchange membrane (PEM) fuel cells, alkaline fuel cells, phosphoric acid fuel cells, molten carbonate fuel cells, and solid oxide fuel cells).

Recently, electrospinning technique has been induced into the fuel cells for fabricating high efficient catalysts and proton exchange membranes.

4.6.1 Electrospun Nanofiber-Based Anode Catalyst

Anode catalyst is the key component in Fuel cell. Usually, high platinum (Pt) content is needed as anode catalyst in a thin layer to minimize the ohmic resistance and mass transfer limitations. For the high expense of Pt, many researchers began to reduce the amount of Pt within the fuel cell; thus, high efficient catalysts and catalyst supports with large surface area are required. For this purpose, two main routes have been used: **Route 1**: Improve the surface areas of Pt nanostructures. For example, Kim et al. [23] fabricated electrospun PtRu, PtRh, Pt/carbon and Pt nanowires as catalyst, which exhibited better performance than conventional carbon-supported Pt or Pt black for the enhanced electrical properties and pertinent interface formation based on 1D nanostructures. **Route 2:** Disperse catalysts uniformly on porous support materials. For example, Xia and co-workers [20] fabricated Pd nanoparticles or nanowires on electrospun TiO_2 nanofibers, which exhibited higher electrochemical activity and durability for direct methanol oxidation compared with commercial Pt/C catalyst. *Details of the electrospun nanofiber-based anode catalysts in the fuel cells will not be illustrated here for they have been illustrated in the section of catalysts.*

4.6.2 Electrospun Nanofiber-Based Proton Exchange Membranes

The proton exchange membrane is the central element of a PEM fuel cell. To function, the membrane must absorb water and conduct protons, but not electrons or gas with sufficient mechanical strength. Traditionally, the perfluorosulfonated membranes such as Nafion have been widely used for their excellent oxidative and chemical stability as well as high proton conductivity. From 2008, a series of proton exchange membranes containing electrospun-able polymer/Nafion composite fibers have been fabricated as the proton exchange membranes such as polyacrylic acid (PAA)/Nafion, polyethylene oxide (PEO)/Nafion, polyvinyl alcohol (PVA)/Nafion, and polyvinyl pyrrolidone (PVP)/Nafion composite fibers [74–77]. Among those composite fibers, PVP/alcohol had the best compatibility with Nafion/Nafion-PTFE with smooth surface. In 2010, Kawakami and Tamura [78] firstly investigated the proton-exchanging performances based on electrospun fiber-based proton exchange membrane containing aligned sulfonated polyimide nanofibers and sulfonated polyimide. Dramatically enhanced proton conductivity, lower gas permeability, and longer durability have been obtained. Recently, a heterogeneous nonwoven with sandwich structure by program-controlled co-electrospinning has been demonstrated by Liu and Meng [79]. Firstly, sulfonated poly (fluorenyl ether ketone) (SPFEK) was electrospun for the first layer. Then, the gap-filling materials polyether sulfone (PES) and SPFEK were electrospun simultaneously. Consequently, SPFEK nanofiber was electrospun onto the existed two layers. Finally, the in situ gap filling can be achieved by dissolving PES followed by dry as shown in Fig. 4.41. Such membrane possessed excellent oxidative stability and a high proton conductivity of 0.0056–0.0061 S cm^{-1} at 30–80 °C with an extremely low activation energy of 1.30 kJ/mol.

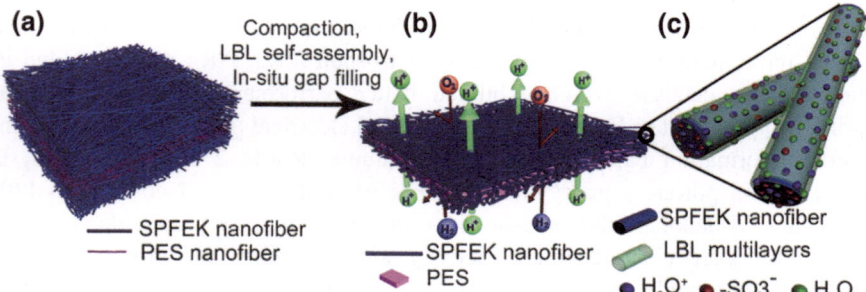

Fig. 4.41 Schematic illustration of sandwich-structured nanofiber-based proton-exchanging membranes. Reproduced from Ref. [79] by permission of The Royal Society of Chemistry

4.6.3 Future Challenge

In principle, electrospun nanofibers-based catalysts and proton exchange membranes offer the good opportunity for high efficient fuel cells driven by their large surface-to-volume rate and controlled diversity in compositions. But several important issues should be clear:

1. Little attention has been paid to investigate the relationship between the nature (diameter, size of gains, pores, etc.) of single nanofibers and the performances of the whole fiber membranes.
2. Lack of systematically theoretical exploration.

4.7 Biomedical Field

Electrospun nanofibers have multitude of possible applications in medicine and pharmacy for proteins, viruses, and bacteria have dimensions on this order. In this section, some current activities will be addressed.

4.7.1 Drug Delivery

Usually, drug delivery systems are used to improve therapeutic efficacy and safety of drugs by delivering them over a period of treatment to the site of action at a rate dictated by the need of physiological environment. Electrospun nanofibrous scaffolds have been used as carriers for diverse types of drugs, genes, and growth factors, for one decade, whereby the release profile can be finely controlled by modulation of the scaffold's morphology, porosity, and composition. Electrospun scaffolds as drug carriers are promising for future biomedical applications, particularly in the prevention of postsurgical adhesions and infections, for postoperative local chemotherapy, and for bone and skin tissue engineering owing to those electrospun scaffold can offer site-specific delivery of drugs to the body. In 2002, Wnek and groups [80] firstly explored the electrospun fiber mats as drug delivery vehicles with tetracycline hydrochloride as a model drug. In their study, electrospun fiber mats made from polylactic acid (PLA), poly(ethylene-co-vinyl acetate) (PEVA), or from a 50:50 blend of the two were electrospun from chloroform solutions containing a small amount of methanol to solubilize the drug (Fig. 4.42a and b). The release of the drug from the drug delivery systems was traced by UV–vis spectroscopy (c and d) in contrast to the commercially drug delivery system, Actisite (Alza Corporation, Palo Alto, CA) as shown in Fig. 40, as well as to cast films of the various formulations. Both electrospun PEVA and 50/50 PLA/PEVA mats gave relatively smooth release of drug over about 5 days.

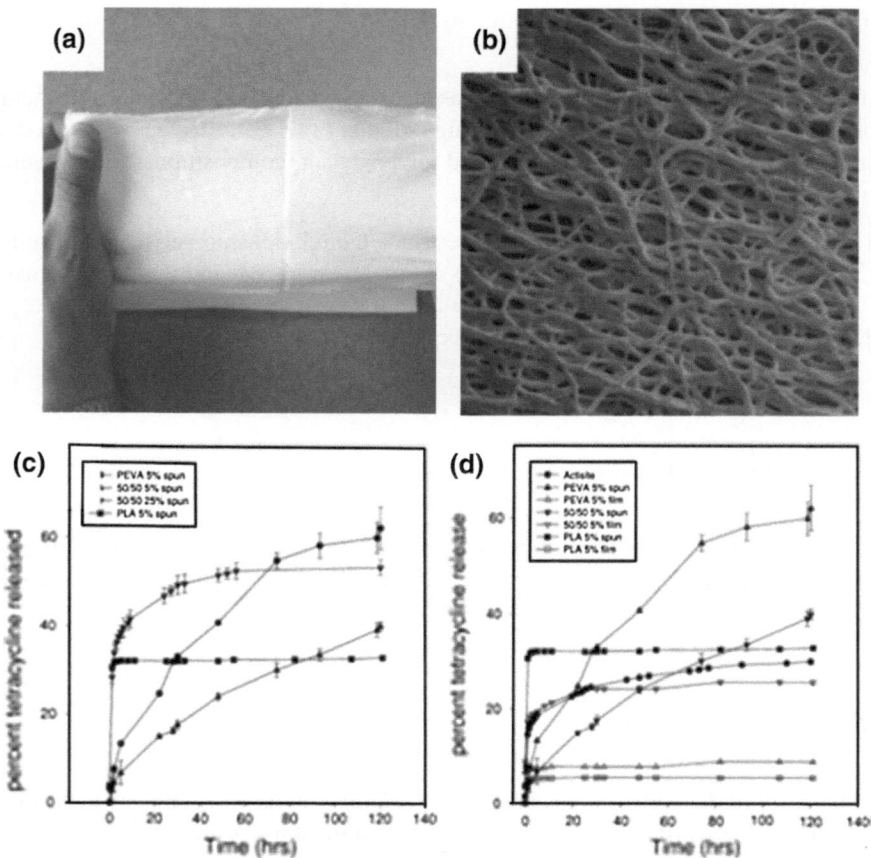

Fig. 4.42 Photograp **a** and SEM **b** of the PEVA electrospun mats. **c** Percentage release of tetracycline HCl from electrospun mats versus time. **d** Percentage release of tetracycline HCl from films and electrospun mats versus time. Reprinted from Ref. [80], Copyright 2002, with permission from Elsevier

Prompted by this work, four main types of drug delivery systems based on electrospun fiber mats have been developed such antibiotic drug delivery system, anticancer drug delivery system, gene delivery system, and biomolecule delivery systems.

4.7.1.1 Antibiotic Drug Delivery System

In 2004, Hsiao and Chu [81] investigated the incorporation and controlled release of hydrophilic cefoxitin sodium (MeforxinR, an antibiotic drug) based on electrospun poly(lactide-co-glycolide) fiber mats. The drug can be successfully incorporated and released from electrospun fiber mats without loss of structure and

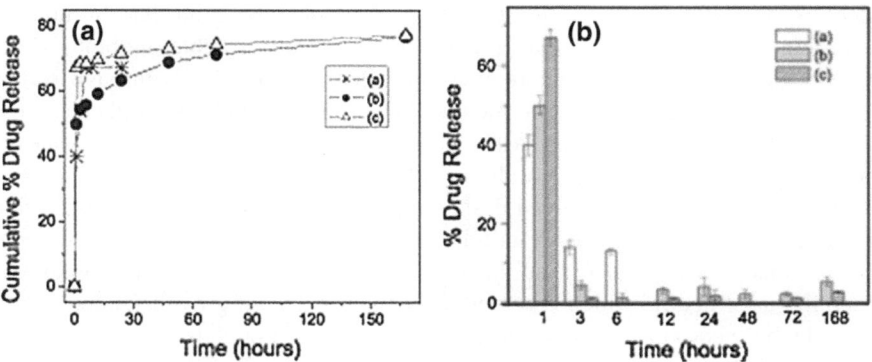

Fig. 4.43 Cumulative curve **a** and differential curve **b** of the drug release profile from medicated electrospun mats. *a*: medicated PLGA with 1 wt % drug, *b*: medicated PLGA/PEG-b-PLA blend with 5 wt % drug, and *c*: medicated PLGA with 5 wt % drug. Reprinted from Ref. [81], Copyright 2004, with permission from Elsevier

bioactivity. The drug release behavior from the electrospun fiber mats (Fig. 4.43) and its antimicrobial effect on Staphylococcus aureus cultures were investigated (Fig. 4.44). Additionally, amphiphilic block copolymer (PEG-b-PLA) was used to reduce the cumulative amount of the released drug at earlier time points and prolonged the drug release rate at longer time.

It is well known that localized and temporally controlled delivery systems are crucial in achieving high local bioactivity and low systemic side effects of antibiotics in the treatment of dental, periodontal, and bone infections. For this intent, Ma and Feng [82] demonstrated a 3D porous tissue engineering scaffold based on

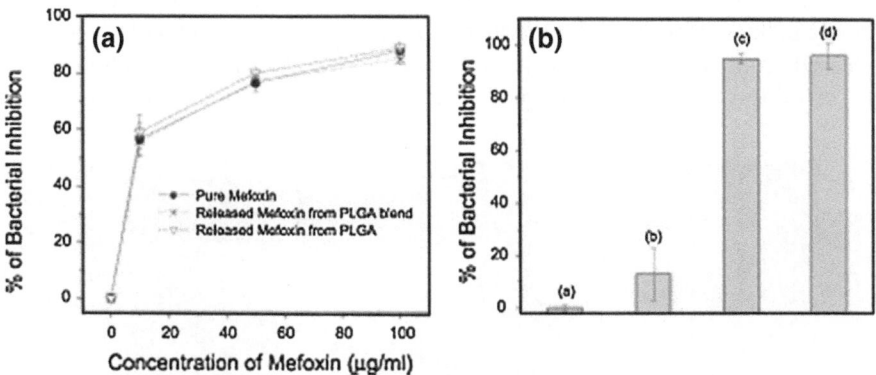

Fig. 4.44 (**a**) and (**b**) are inhibition of bacterial growth after 24 h of incubation as a function of the concentration of pure or released cefoxitin sodium, expressed as percentage of initial inoculation amount. *a* : control (without drug and scaffold), *b*: PLGA/PLA/PEG-b-PLA scaffold without drug, *c*: medicated PLGA/PLA/PEG-b-PLA scaffold having 1.59 mg of cefoxitin sodium, and *d*: medicated PLGA scaffold having 1.64 mg of cefoxitin sodium. Reprinted from Ref. [81], Copyright 2004, with permission from Elsevier

electrospun fiber with the ability to release antibiotics in a controlled fashion for long-term inhibition of bacterial growth. In their study, soluble antibiotic drug, doxycycline (DOXY), was incorporated into PLGA nanospheres through water-in-oil-in-oil (W/O/O) emulsion method. Then, the PLGA nanospheres were incorporated into prefabricated PLLA fiber mats with well-interconnected macro-porous structure (Fig. 4.45). Results of in vitro release and antibacterial experiments suggest that the developed drug-containing nanofibrous scaffold is capable of effectively delivering DOXY in a controlled fashion with prolonged duration. Those biodegradable PLLA scaffolds have well-interconnected macro-porous and nanofibrous structure, which can inhibit common bacterial growth for more than 6 weeks by incorporating DOXY/PLGA nanospheres (Fig. 4.46).

4.7.1.2 Anticancer Drug Delivery System

In 2005, Jing and groups [83] incorporated anticancer drug (doxorubicin hydrochloride: Dox) into amphiphilic poly(ethylene glycol)-poly(L-lactic acid) (PEG-LLA) diblock copolymer nanofiber mats via water-in-oil (emulsions) electrospinning, in which aqueous phase is for drug and oil phase is for diblock copolymer. The whole steps can be found in Fig. 4.47a. The released Dox was monitored through UV absorbance of Dox at 483.5 nm as a function of incubation time. Without proteinase K, the drug release behaviors are closely related to the distribution status of the drug (Fig. 4.47b). An approximately linear relationship between the fraction of released drug (M_t/M_∞) and $t^{1/2}$ for the first 75 % of drug release was obtained, meaning the Dox release obeys the diffusion mechanism. Additionally, the authors also found that the fast drug release rate can be obtained by increasing the PEG block length within the whole diblock copolymer (Fig. 4.47d and e), meaning the enzymatic degradation is predominant during Dox release in the second stage. Furthermore, the authors also proven that Dox could be released from the fibers without losing cytotoxicity against mice glioma cells (C_6 cell lines) through MTT method (Fig. 4.48). Following this work, some other

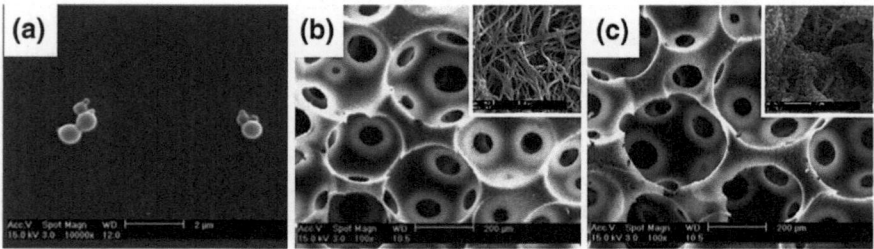

Fig. 4.45 a SEM image of the PLGA85-142 K nanospheres containing DOXY. **b** and **c** SEM images of the PLLA scaffold before and after PLGA nanospheres, respectively, with low magnification. The insets in **a** and (**c**) are SEM images with high magnification. PLGA85-142 K means ratio of LA/GA is 85–15 with the molecule weight of 142 kDa and sphere diameter of 730 $\pm$ 160 nm. Reprinted from Ref. [82], Copyright 2010, with permission from Elsevier

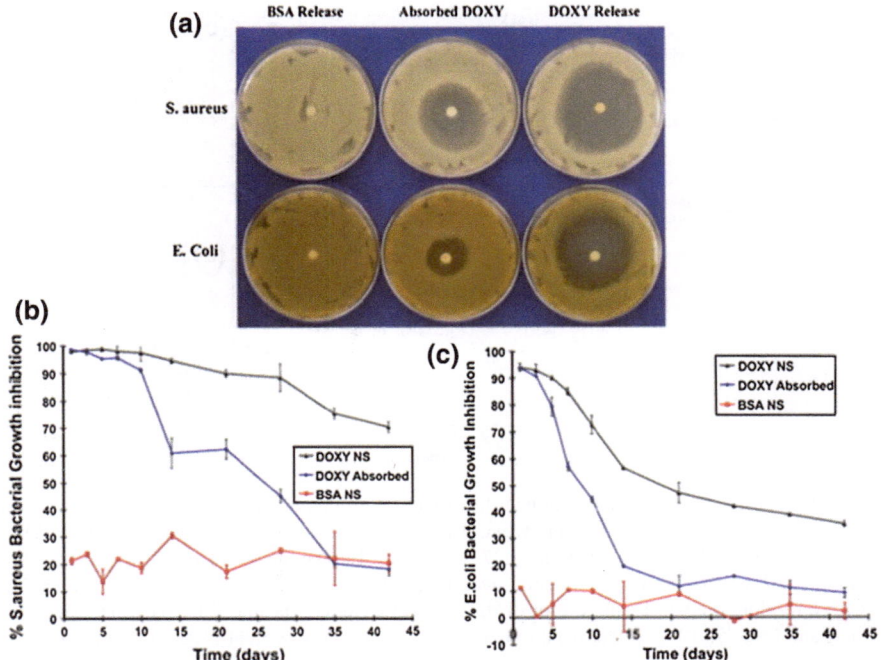

Fig. 4.46 a Agar petri dish cultivated with *S. aureus* (*upper*) and *E. coli* (*lower*) with scaffold samples in the center after 5 days of incubation. Long-term *S. aureus* **b** and *E. coli* **c** growth inhibition test for release solution from three groups of scaffolds. Reprinted from Ref. [82], Copyright 2010, with permission from Elsevier

anticancer drugs have been successfully incorporated into the electrospun fiber scaffold such as [1, 3-bis(2-chloroethyl)-1-nitrosourea](BCNU) into PEG-PLLA fiber scaffold [84], paclitaxel into PLGA fiber scaffold [85] [86–92].

For effective cancer gene therapy, tumor-targeting adenoviral (Ad) complexes are critical for delivery to both primary and metastatic lesions. In 2010, Kim and Yun [93] fabricated nanocomplexes of Ad, chitosan, PEG, and folic acid (FA) for effective FA-receptor-expressing tumor-specific transduction. In their study, chitosan–PEG–FA conjugates were generated and processed by electrospinning for adenoviral encapsulation via in situ ionic cross-linking with tripolyphosphate (TPP) (Scheme 4.48). Ad/chitosan nanocomplexes retained biological activity without impairment of the transduction efficiency of naked Ad. Chitosan layer on the adenoviral surface provided chemical conjugation sites for PEG and further for FA, as a targeting moiety at the end of heterofunctional PEG. In addition, the transduction efficiency of Ad/chitosan–PEG–FA was proportionally increased with the amount of surface-bound FA on Ad nanocomplexes, proving the FA-mediated viral entry. GFP gene expression transduced by tumor-specific FA-mediated Ad/chitosan–PEG–FA nanocomplexes was about 57.2 % higher than that obtained by nonspecifically transducing Ad/chitosan nanocomplexes confirming that

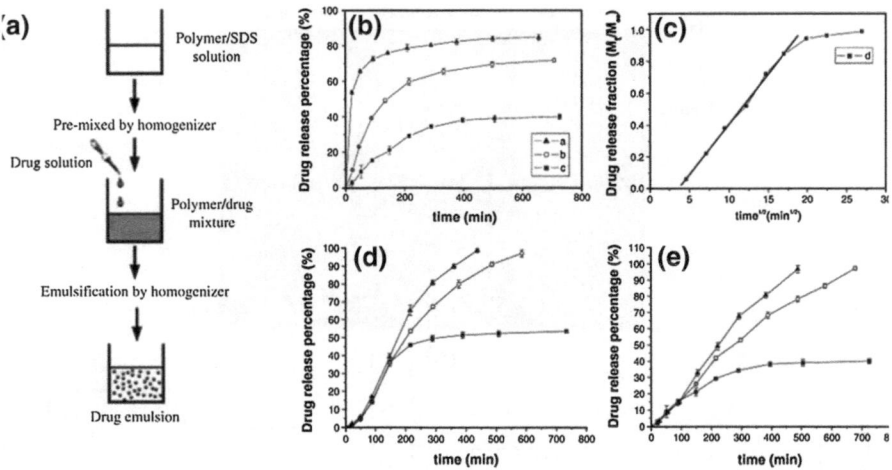

Fig. 4.47 a Preparation of W/O emulsions. **b** Release profiles of Dox from electrospun fibers in 0.05 mol/l Tris–HCl buffer solutions without proteinase K at 37 C: (*a*) Suspension-spun Dox/PEG5000-PLLA fibers, Dox content 2.27 wt %; (*b*) Emulsion-spun Dox/PLLA fibers, Dox content 2.10 wt %; (*c*) Emulsion-spun Dox/PEG750-PLLA fibers, Dox content 2.06 wt %. **c** Curve re-plotted against square root of time. **d** and **e** Release profiles of Dox from PEG5000-PLLA and PEG750-PLLA electrospun fibers, respectively. Dox content 2.12 wt %. Proteinase K concentration in the buffer: (■) none; (○) 3 μg/ml; (▲) 5 μg/ml. Reprinted from Ref. [83], Copyright 2005, with permission from Elsevier

infection through receptor-mediated endocytosis is more efficient than nonspecific charge interaction-mediated infection (Fig. 4.49).

It is well known that the drug release of resultant drug-loaded fibers not only occurs via diffusion, but also through loading material degradation. Thus, the rate of material degradation can influence the drug release prolife as well. Jing and co-workers [94, 95] published a series of papers in this field. They investigated the influences of surfactants and anticancer drugs on the diameter size and uniformity

Fig. 4.48 In vitro cytotoxicity of Dox-loaded PEG-PLLA fibers against mice glioma cells (C$_6$ cell lines). The general test procedure is described in the text. Cell density: 1×10^4 cells/ml. Reprinted from Ref. [83], Copyright 2005, with permission from Elsevier

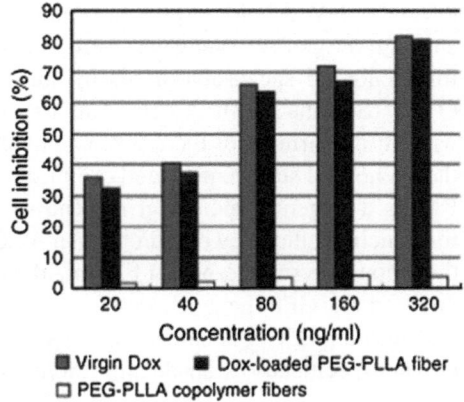

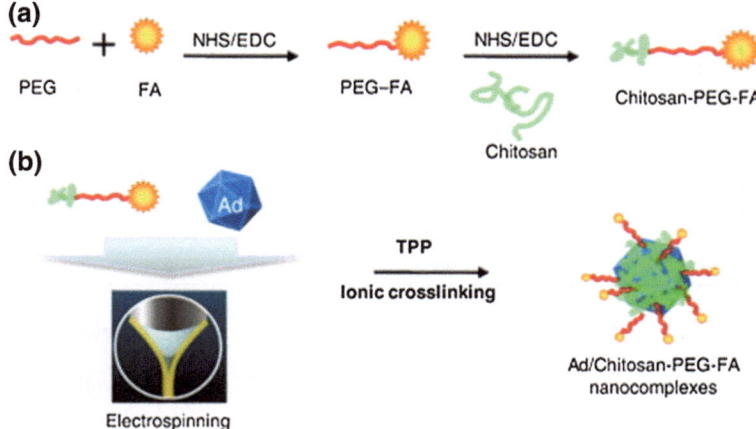

Fig. 4.49 Schematic of chitosan–PEG–FA synthesis (**a**) and electrospinning process for generating Ad/chitosan–PEG–FA nanocomplexes (**b**). Reprinted from Ref. [93], Copyright 2010, with permission from Elsevier

of electrospun PLLA fibers by adding various surfactants (cationic, anionic, and nonionic) and anticancer drugs into the PLLA solution. Additionally, they also tested the influences of surfactants and anticancer drugs on the diameter size and uniformity of electrospun PLLA fibers by adding various surfactants (cationic, anionic, and nonionic) and anticancer drugs into the PLLA solution. Those results indicated that the drug release profile of doxorubicin-hydrochloride-embedded PLLA electrospun fibers obeyed nearly zero-order kinetics in the presence of proteinase K due to the degradation of PLLA fibers. Namely, the degradation of PLLA played a key role in the drug release.

4.7.1.3 Gene Delivery System

Scaffold-mediated gene delivery or reverse transfection, in the form of hydrogels or porous scaffolding materials, has been studied for DNA delivery for long time through gas-forming or particle-leaching techniques. Recently, nanofibrous scaffolds represent a novel class of potent materials for such applications gain special attention for they can provide a good representation of the nanofibrous architecture of the natural extracellular matrix, which can provide an extra dimension for better control over cellular functions.

Extracellular and intracellular barriers typically prevent nonviral gene vectors from having effective transfection efficiency. Formulation of a gene delivery vehicle that can overcome the barriers is a key step for successful tissue regeneration. In 2005, Liang and Chu [96] developed a novel core–shelled DNA nanoparticle by invoking solvent-induced condensation of pDNA (b-galactosidase or GFP) in a solvent mixture (94 % DMF + 6 % Tris/EDTA buffer) and subsequent encapsulation of the condensed DNA globule in a triblock copolymer,

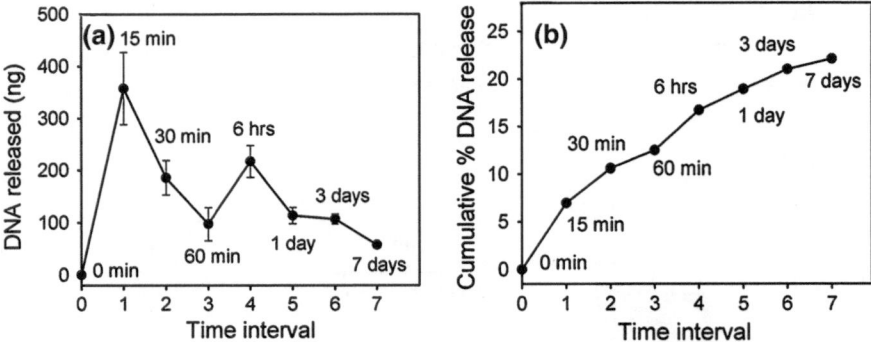

Fig. 4.50 a and **b** Shows the controlled release of DNA from PLGA scaffold with poly(lactide)-b-poly(ethylene glycol)-b-poly(lactide) (LEL). Reprinted from Ref. [96]. Copyright 2005, Oxford University Press

poly(lactide)-b-poly(ethylene glycol)-b-poly(lactide) (PLLA–PEG–PLLA: LEL). The bioactive pDNA was found to be released in an intact form from the scaffold with a controlled release rate and transfect cells in vitro (Fig. 4.50).

Although viral gene transfer is efficient in achieving transgene expression for tissue engineering, drawbacks of virus dissemination, toxicity, and transient gene expression due to immune response have hindered its widespread application. Many tissue engineering studies thus opt to genetically engineer cells in vitro prior to their introduction in vivo. However, it would be attractive to obviate the need for in vitro manipulation by transducing the infiltrating progenitor cells in situ. Recently, Liao and Leong [97] fabricated virus-encapsulating electrospun fibrous scaffold for sustained and localized transduction. In their study, adenovirus encoding the gene for green fluorescent protein was incorporated into the core of PCL fibers through coaxial electrospinning and released in a porogen-assisted manner. Cells seeded on the virus-encapsulating scaffold exhibited transgene expression more than one month with a reduced proliferation rate (Fig. 4.51). RAW 264.7 cells cultured on the virus-encapsulating fibers produced a lower level of interleukin-1β, tumor necrosis factor-α and interferon-α, confirming the reduced activation of macrophage cells (Fig. 4.52).

It should be pointed out that Saraf and Mikos [98] described the use of coaxial electrospinning as a means for the fabrication of fiber mesh scaffolds and for the encapsulation and subsequent release of a nonviral gene delivery vector over a period of up to 60 days (Fig. 4.49) (Fig. 4.53).

4.7.1.4 Biomolecule Delivery System

Thus far, two routes [99] have been regarded as the most promising tissue engineering methods: **Route 1:** Implant pre-cultured cells and synthetic scaffold complexes into the defect place in which the seeded cells are generally isolated from host target tissues, providing the main resource to form newly grown tissue.

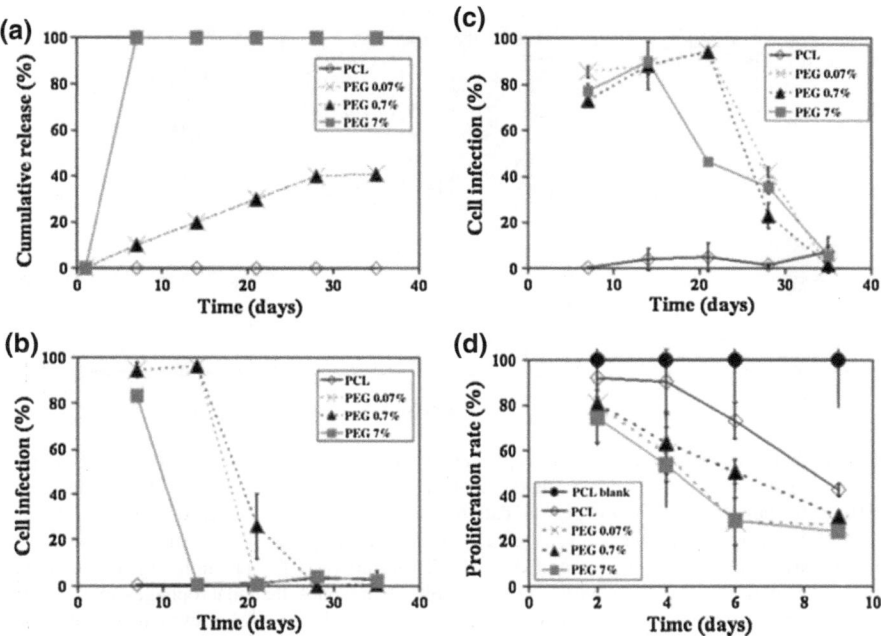

Fig. 4.51 Infection of cells through virus-encapsulated scaffolds. Infection of HEK 293 cells by GFP-adenovirus-encapsulated fibers produced with different formulations (PCL, PCL/PEG 0.07 %, PCL/PEG 0.7 % and PCL/PEG 7 %). End point dilution assay **a** of the scaffold supernatant reveals that the viruses are continuously released in the first two weeks from the PEG-incorporated fibers but not from the nonporous *PCL* fibers. HEK 293 incubated in scaffold supernatant **b** shows transgene expression only in the first two weeks while cells seeded on the virus-encapsulated scaffolds **c** remain positive over 30 days. **d** Cells cultured on the virus-encapsulated scaffolds show lower proliferation compared to *PCL* scaffold without virus. Cell transgene expression were characterized using flow cytometry and each point represents the mean ±S.D. (*n* = 3). Reprinted from Ref. [97], Copyright 2009, with permission from Elsevier

The synthetic scaffolds provide porous 3D structures to accommodate the cells to form extracellular matrix (ECMs) and regulate the cell growth in vivo. These synthetic scaffolds are biodegradable and degrade in accordance with the tissue regeneration time frame [100, 101]. **Route 2:** Place acellular scaffolds immediately after injury. The key of this approach is using scaffolds to deliver appropriate biomolecules to the defect area; the biomolecules can be released from the scaffolds in a controlled manner and recruit progenitor cells toward the defect area, promoting their proliferation and differentiation and enhancing tissue regeneration. Currently, the main focus lies in combining these two approaches to taking the advantages of the scaffolds with controlled release of biomolecules for seeded cells to proliferate and differentiate during an ex vivo pre-culture period, thereby encouraging tissue formation after implantation in vivo [102–105].

Recently, electrospinning has gained exponentially increasing popularity for the preparation of bioactive scaffolds with biomolecule release due to their structural

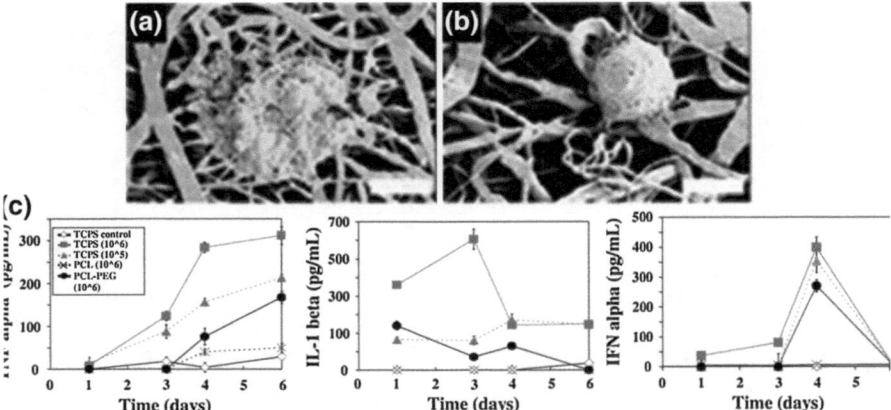

Fig. 4.52 Reduced activation of macrophages by sustained delivery of viral vectors. Culture of macrophage cells on virus-encapsulated fibers (10^6 IFU/PFU/mL/scaffold, PCL or PCL/PEG 0.07 % formulations). SEM images indicate that macrophage cells seeded on PEG-incorporated and virus-encapsulated fibers show an activated cell morphology **a** while cells seeded on nonporous PCL fibers remain rounded **b**. **c** ELISA analysis for pro-inflammatory cytokines (TNF-α and IL-1 β) and anti-viral cytokine (IFN-α) indicates that macrophages exposed to a sustained release of the viral vectors (PCL-PEG) produce reduced level of cytokines compared to a direct bolus exposure (TCPS 10^5 and 10^6). Macrophage cells cultured on virus-encapsulated but nonporous PCL fibers appear to have no exposure to the virus and produce only baseline level of all three cytokines. Scale bar: 10 μm. Each point represents the mean ±S.D. ($n = 3$). Reprinted from Ref. [97], Copyright 2009, with permission from Elsevier

resemblance to the extracellular matrix (ECM). Li and Loh [106] fabricated hydrogel nanofibers mats from biodegradable and thermoresponsive multi-block poly(ester urethane)s comprising poly(ethylene glycol) (PEG), poly(propylene glycol) (PPG), and poly(ε-caprolactone) (PCL) segments. Such nanofiber mats were more water absorbent under cold conditions and shrunk when exposed to higher temperatures. Namely, the rate of protein release could be controlled by

Fig. 4.53 Cumulative release of rhodamine-tagged *PEI-HA* (r-PEI-HA) from electrospun coaxial fiber meshes at 37 °C in PBS with agitation at 115 rpm. Error bars represent standard deviation for n = 4. Reprinted from Ref. [98], Copyright 2010, with permission from Elsevier

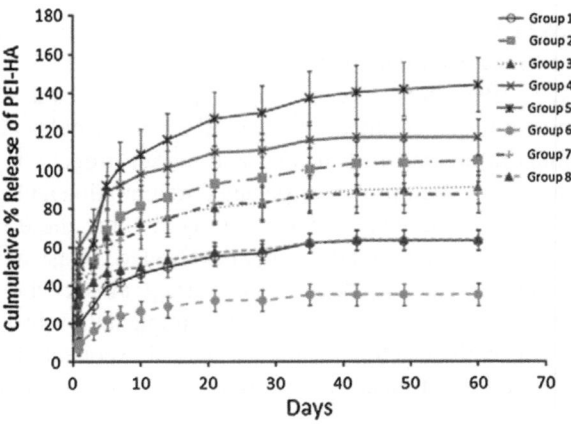

changing the temperature of the nanofiber environment (Fig. 4.54). Furthermore, such hydrogel nanofibers degrade much faster than the PCL nanofibers with small molecular weight of between 1,000 and 4,000, facilitating the removal of these nanofibers after use through renal filtration (Fig. 4.55).

The treatment of challenging fractures and large osseous defects is a formidable problem for orthopedic surgeons. Tissue engineering and regenerative medicine approaches are used to solve this problem by delivering osteogenic signals within scaffolding biomaterials. Kolambkar and Guldberg [107] demonstrated a hybrid growth factor delivery system that consists of an electrospun nanofiber mesh tube for guiding bone regeneration combined with peptide-modified alginate hydrogel injected inside the tube for sustained growth factor release. The ability of this

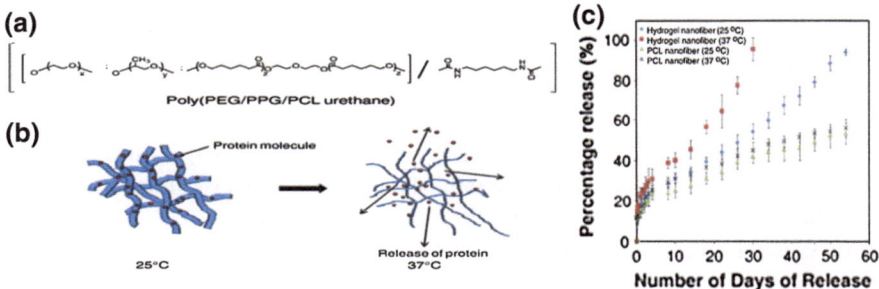

Fig. 4.54 a Structure of the copolymer used in this work. **b** Schematic depiction of the temperature-controlled protein release. **c** Comparing the protein release of the hydrogel nanofibers and *PCL* nanofibers at 25 and 37 °C. Reprinted from Ref. [106], Copyright 2010, with permission from Elsevier

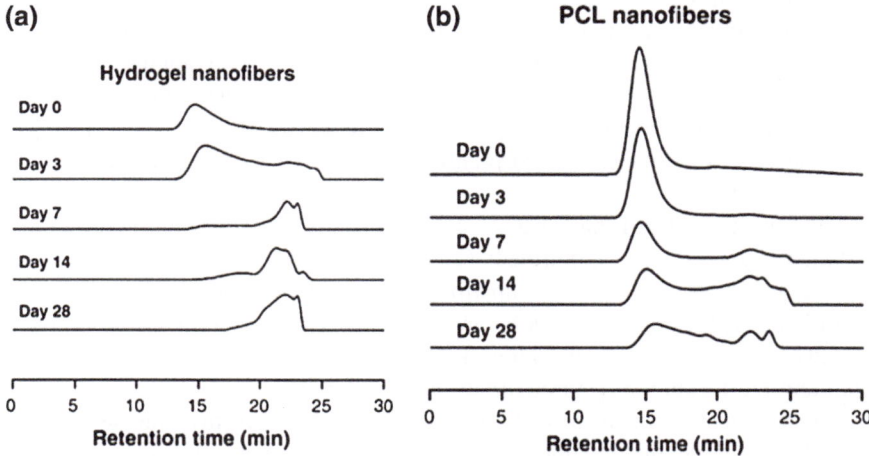

Fig. 4.55 GPC profiles comparing the hydrolytic degradation of hydrogel nanofibers (**a**) and PCL nanofibers (**b**) carried out at pH 11.5 at 37 °C. Reprinted from Ref. [106], Copyright 2010, with permission from Elsevier

system to deliver recombinant bone morphogenetic protein-2 (rhBMP-2) for the repair of critically sized segmental bone defects in a rat model was tested. Longitudinal m-CT analysis and torsional testing provided quantitative assessment of bone regeneration. The authors found that the hybrid delivery system resulted in consistent bony bridging of the challenging bone defects. However, in the absence of rhBMP-2, the use of nanofiber mesh tube and alginate did not result in substantial bone formation. Perforations in the nanofiber mesh accelerated the rhBMP-2-mediated bone repair and resulted in functional restoration of the regenerated bone. m-CT-based angiography indicated that perforations did not significantly affect the revascularization of defects, suggesting that some other interaction with the tissue surrounding the defect such as improved infiltration of osteoprogenitor cells contributed to the observed differences in repair.

4.7.2 Tissue Engineering

For many exhaustive reviews [108–124] and books [125–149] on TE based on electrospinning technique have been published, herein a concise illustration of the activities and research directions in TE will be presented.

In general, tissue engineering is a multi-disciplinary field to develop and manufacture therapeutic products that utilize the combination of matrix scaffolds with viable human cell systems or cell responsive biomolecules derived from cells, for the repair, restoration, or regeneration of cells or tissue damaged by injury, disease, or congenital defects. Tissue engineering (TE) involve scaffolds or matrices to provide support for cells in order to express new ECM. The biocompatibility of scaffold materials actively participates in the signaling process for the requirement of safe degradation and also provides a substrate for cell migration into the defect sites of the tissue. It has been proven that scaffolds mimicking the architecture of the extracellular matrix should offer great advantages for tissue. Thus, an ideal scaffold should replicate the structure and function of the natural extracellular matrix as closely as possible, until the seeded cells have formed a new matrix, which makes electrospinning prevails as the fabrication method for these scaffolds.

Till now, diverse cells have been seeded onto carrier matrices for the generation of target tissue. Results confirmed: (a) Fiber architecture generally affects cell growth positively; (b) The biocompatibility of a material improves by decreasing fiber diameter; (c) Porosity has a favorable influence on cell growth (the degree of porosity and the average pore dimensions are significant factors for cell proliferation and the formation of three-dimensional tissues).

Usually, electrospun fiber scaffolds do not have enough mechanical compatibility. For example, the maximum deformation of cartilage is 20–120 %. However, maximum elongation based on electrospun collagen fiber is usually less than 10 %.

Totally, scaffolds fabricated from electrospun nanofibers have several clear advantages. However, considerable room for optimization remains with respect to architecture, surface properties, biodegradability, porosity, and mechanical properties and also with respect to the seeding of cells in the three-dimensional space and the supply of nutrients to the cells. It is often observed that the cells preferentially grow on the surfaces or that they initially adhere to the carrier fibers, but then detach after differentiation (Fig. 4.56).

4.8 Other Applications

4.8.1 Template

Electrospun polymeric fibers can act as template for 1D nanostructures. In the template process, electrospun polymeric nanofibers are usually coated with or blended with polymer, metal precursor, and other precursors. Then, the template fibers can be selectively removed by extraction, degradation, and calcination. For example, Greiner and co-workers [150] demonstrated the polymer, metal, and hybrid nano-and mesotubes by coating degradable polymer template fibers (TUFT Process). In their study, degradable polymer [Poly(L-lactide) (PLA)] electrospun fibers were used as the templates. Then, hollow polymer, metal, and hybrid nano-and mesotubes can be obtained by coating them or the corresponding precursors on

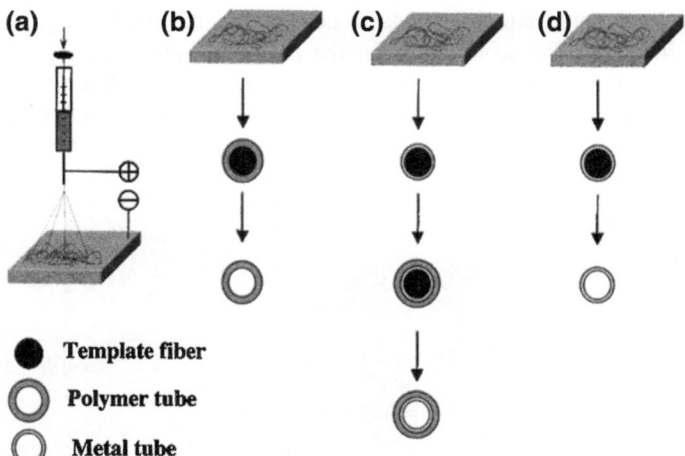

Fig. 4.56 a Schematic diagram of the apparatus used for electrospinning. It consists of a syringe with a metal capillary (diameter 0.3 mm) and a pressure supply on the piston of the syringe. A high-voltage field, typically in the range of 4 kV/cm, with the anode on the metal capillary, was applied to the polymer solution. Polymer fibers are formed as soon as a drop of the polymer solution has been formed at the tip of the capillary. These fibers are deposited on a glass substrate. **b** Concept for the preparation of polymer tube. **c** Concept for the preparation of polymer/metal hybrid tubes. **d** Concept for the preparation of metal tubes. Reproduced from Ref. [150] by permission of John Wiley and Sons Ltd

the outer surface of the PLA. Finally, PLA fiber can be removed by the post-treatment and hollow tubes can be obtained as shown in Scheme 11. In this paper, hollow poly(p-xylylene) (PPX) tubes (Fig. 4.57a) and Al tubes (Fig. 4.57b) can be successfully obtained. Following this work, a lot of 1D tubes have been successfully obtained.

In 2003, Xia and Li [151] demonstrated a novel usage of electrospun polymeric template. In their study, the precursors of the titania (titanium tetraisopropoxide: Ti $(OiPr)_4$) were blended with PVP in the solution. Then, the composites were electrospun to form the composite fibers. Finally, PVP was selectively removed by sintering the composite fiber 500 °C for 20 h in air and titanium tetraisopropoxide was converted into crystal titania nanofibers as shown in Fig. 4.58. Following this work, a lot of metal or metal oxide fibers have been successfully achieved.

4.8.2 Water Treatment

4.8.2.1 Heavy Metal Ion Adsorption

Heavy metals are a serious biological problem in aquatic systems. Adsorption and filtration are the commonly used methods for removal of these contaminants. Recently, electrospun silk nanofiber mats [152] have been applied in removing heavy metal ions. Higher adsorption capacity (1.65–2.88 mg/g) was obtained based on electrospun nanofiber mats in contrast to that (0.71 mg/g) of the conventional materials like wool silver for the larger surface area of the fiber mats. In addition to those polymeric nanofiber mats, metal oxide nanofiber mats have also applied in heavy metal adsorption. For example, Wang and Li [153] used

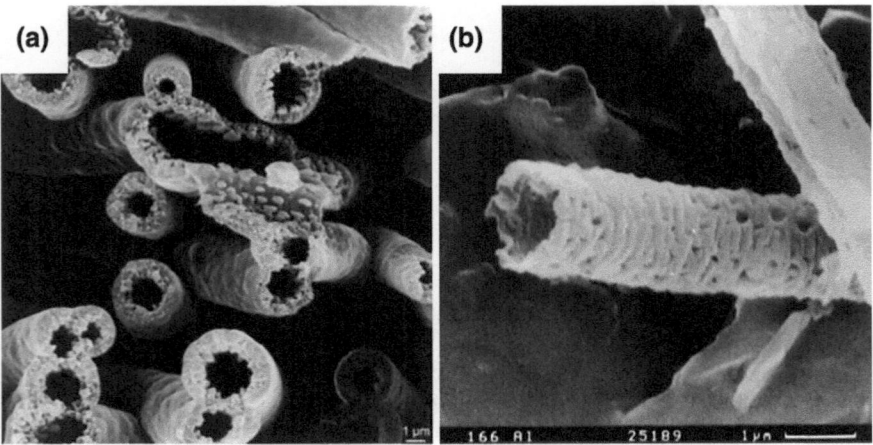

Fig. 4.57 SEM images of PPX (**a**) and Al (**b**) hollow tubes via PLA fiber as template. Reproduced from Ref. [150] by permission of John Wiley and Sons Ltd

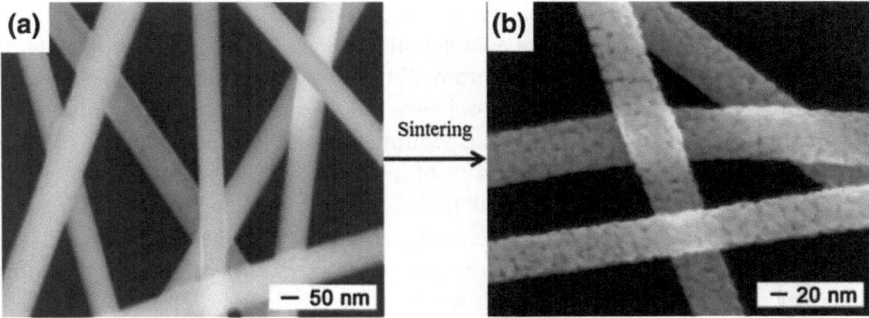

Fig. 4.58 SEM images of the titanium tetraisopropoxide/PVP composite fibers (**a**) and crystal titania nanofibers (**b**), respectively. Reprinted with the permission from Ref. [151]. Copyright 2003 American Chemical Society

anatase mesoporous titanium nanofiber mats for the adsorption of Cu (II) ions. The adsorption capacity can reach 12.8 ± 0.6 mg/g.

4.8.2.2 Adsorption of Organic Compounds

Organic materials in drinking water can pose health hazards and also need to be removed. Kaur et al. [154] explored the removal of phenolphthalein as a model organic molecule from water using a poly(methylmethacrylate) (PMMA) nanofiber membrane functionalized with phenylcarbomylated and azidophenylcarbomylated β-cyclodextrins. The results obtained showed that the functionalized nanofibrous membranes were able to effectively capture the PHP molecules.

Metal oxide fiber mats also exhibited good ability in organic compounds removal from water for its filtration and photocatalytic ability. Zhang and Sun [155] used titanium oxide nanofiber mats to remove humic acid in water. In their study, TiO_2 nanowire membrane achieved near 100 and 93.6 % removal rate of humic acid and total organic carbon (TOC), respectively, via a concurrent filtration and photocatalytic oxidation. The TiO_2 nanowire membrane also showed excellent antifouling ability owing to the photodegradation of foulants by the TiO_2 nanowire membrane.

4.8.3 Biosensors

Electrospun products gain special attention for sensitive detection of clinical, environmental, and food safety analytes driven by their extremely large surface area for providing a number of binding sites available for biological recognition element immobilization and diverse functionalization. Currently, electrospun nanofibers, nanowires, and nanotubes are frequently investigated for use in biosensors. For example, Ranaei-Siadat and group [156] created poly(vinyl alcohol) (PVA) nanofibers containing acetylcholinesterase (AChE) by electrospinning a melt of PVA, AChE, and bovine serum albumen (BSA) as an enzyme stabilizer. The AChE-modified nanofibers exhibited a 40 % activity recovery after electrospinning. Additionally, the enzymes within the nanofibers had a higher stability (the immobilized AChE retained >34 % of its initial activity when stored at 30 °C for 100 days and retained 70 % of its initial activity after ten consecutive reactor batch cycles) in acidic solutions when compared to free enzymes.

Recently, the incorporation of molecularly imprinted polymers (MIPs) within nanofiber networks to construct high-sensitivity analytical systems has also been developed. For example, Haupt and Gheber [157] fabricated PVA nanofibers to support MIP (imprinted with the fluorescent amino acid derivative dansyl-L-phenylalanine) nanoparticles. Kim and Chang [158] fabricated polyimide nanofibers imprinted using a diamine monomer template for binding and detecting estrone.

Wang and Yu [159] reported a facile route to fabricate gold nanoparticles-poly(vinyl alcohol) (Au NPs-PVA) hybrid water-stable nanofibrous mats with tunable densities of Au NPs and further demonstrate the potential application of as-prepared Au NPs-PVA nanofibrous mats as efficient biosensor for the detection of H_2O_2.

References

1. Chand S (2000) Carbon fibers for composites. J Mater Sci 35(6):1303–1313
2. Bergshoef MM, Vancso GJ (1999) Transparent nanocomposites with ultrathin, electrospun nylon-4, 6 fiber reinforcement. Adv Mater 11(16):1362–1365. doi:10.1002/(sici)1521-4095(199911)11:16<1362:aid-adma1362>3.0.co;2-x
3. J-S Kim, Reneker DH (1999) Mechanical properties of composites using ultrafine electrospun fibers. Polym Compos 20(1):124–131. doi:10.1002/pc.10340
4. Dzenis Y (2008) Materials science—structural nanocomposites. Science 319(5862):419–420. doi:10.1126/science.1151434
5. Fong H (2004) Electrospun nylon 6 nanofiber reinforced BIS-GMA/TEGDMA dental restorative composite resins. Polymer 45(7):2427–2432. doi:10.1016/j.polymer.2004.01.067
6. Tian M, Gao Y, Liu Y, Liao Y, Xu R, Hedin NE, Fong H (2007) Bis-GMA/TEGDMA dental composites reinforced with electrospun nylon 6 nanocomposite nanofibers containing highly aligned fibrillar silicate single crystals. Polymer 48(9):2720–2728. doi:10.1016/j.polymer.2007.03.032
7. Gao Y, Sagi S, Zhang L, Liao Y, Cowles DM, Sun Y, Fong H (2008) Electrospun nano-scaled glass fiber reinforcement of bis-GMA/TEGDMA dental composites. J Appl Polym Sci 110(4):2063–2070. doi:10.1002/app.28695
8. Sun W, Cai Q, Li P, Deng X, Wei Y, Xu M, Yang X (2010) Post-draw PAN–PMMA nanofiber reinforced and toughened Bis-GMA dental restorative composite. Dent Mater 26(9):873–880. doi:10.1016/j.dental.2010.03.022
9. Han SO, Son WK, Youk JH, Park WH (2008) Electrospinning of ultrafine cellulose fibers and fabrication of poly (butylene succinate) biocomposites reinforced by them. J Appl Polym Sci 107(3):1954–1959. doi:10.1002/app.26643

10. Tang C, Liu H (2008) Cellulose nanofiber reinforced poly (vinyl alcohol) composite film with high visible light transmittance. Compos Part A: Appl Sci Manuf 39(10):1638–1643. doi:10.1016/j.compositesa.2008.07.005

11. Özden E, Menceloğlu YZ, Papila M (2010) Engineering chemistry of electrospun nanofibers and interfaces in nanocomposites for superior mechanical properties. ACS Appl Mat Interfaces 2(7):1788–1793. doi:10.1021/am100288r

12. Suthat A, Chase G (2001) Nanofibres in filter media. Chem Eng 726:26–28

13. Podgórski A, Bałazy A, Gradoń L (2006) Application of nanofibers to improve the filtration efficiency of the most penetrating aerosol particles in fibrous filters. Chem Eng Sci 61(20):6804–6815. doi:10.1016/j.ces.2006.07.022

14. Ahn YC, Park SK, Kim GT, Hwang YJ, Lee CG, Shin HS, Lee JK (2006) Development of high efficiency nanofilters made of nanofibers. Curr Appl Phys 6(6):1030–1035. doi:10.1016/j.cap.2005.07.013

15. Aussawasathien D, Teerawattananon C, Vongachariya A (2008) Separation of micron to sub-micron particles from water: electrospun nylon-6 nanofibrous membranes as pre-filters. J Membr Sci 315(1–2):11–19. doi:10.1016/j.memsci.2008.01.049

16. Zhang S, Shim WS, Kim J (2009) Design of ultra-fine nonwovens via electrospinning of Nylon 6: spinning parameters and filtration efficiency. Mater Des 30(9):3659–3666. doi:10.1016/j.matdes.2009.02.017

17. Jeong EH, Yang H, Youk JH (2007) Preparation of polyurethane cationomer nanofiber mats for use in antimicrobial nanofilter applications. Mater Lett 61(18):3991–3994. doi:10.1016/j.matlet.2007.01.003

18. Lala NL, Ramaseshan R, Li BJ, Sundarrajan S, Barhate RS, Liu YJ, Ramakrishna S (2007) Fabrication of nanofibers with antimicrobial functionality used as filters: protection against bacterial contaminants. Biotechnol Bioeng 97(6):1357–1365. doi:10.1002/bit.21351

19. Hou H, Reneker DH (2004) Carbon nanotubes on carbon nanofibers: a novel structure based on electrospun polymer nanofibers. Adv Mater 16(1):69–73. doi:10.1002/adma.200306205

20. Formo E, Lee E, Campbell D, Xia Y (2008) Functionalization of electrospun TiO2 nanofibers with Pt nanoparticles and nanowires for catalytic applications. Nano Lett 8(2):668–672. doi:10.1021/nl073163v

21. Formo E, Yavuz MS, Lee EP, Lane L, Xia Y (2009) Functionalization of electrospun ceramic nanofibre membranes with noble-metal nanostructures for catalytic applications. J Mater Chem 19(23):3878–3882

22. Su L, Jia W, Schempf A, Ding Y, Lei Y (2009) Free-standing palladium/polyamide 6 nanofibers for electrooxidation of alcohols in alkaline medium. J Phys Chem 113(36):16174–16180. doi:10.1021/jp905606s

23. Kim HJ, Kim YS, Seo MH, Choi SM, Kim WB (2009) Pt and PtRh nanowire electrocatalysts for cyclohexane-fueled polymer electrolyte membrane fuel cell. Electrochem Commun 11(2):446–449. doi:10.1016/j.elecom.2008.12.027

24. Madhugiri S, Sun B, Smirniotis PG, Ferraris JP, Balkus KJ (2004) Electrospun mesoporous titanium dioxide fibers. Microporous Mesoporous Mater 69(1–2):77–83. doi:10.1016/j.micromeso.2003.12.023

25. Liu RL, Ye HY, Xiong XP, Liu HQ (2010) Fabrication of TiO2/ZnO composite nanofibers by electrospinning and their photocatalytic property. Mater Chem Phys 121(3):432–439. doi:10.1016/j.matchemphys.2010.02.002

26. Lin DD, Wu H, Zhang R, Pan W (2009) Enhanced photocatalysis of electrospun Ag–ZnO heterostructured nanofibers. Chem Mater 21(15):3479–3484. doi:10.1021/cm900225p

27. Wang ZJ, Li ZY, Zhang HN, Wang C (2009) Improved photocatalytic activity of mesoporous ZnO-SnO2 coupled nanofibers. Catal Commun 11(4):257–260. doi:10.1016/j.catcom.2009.10.006

28. Wang Y, Santiago-Aviles JJ (2003) Large negative magnetoresistance and two-dimensional weak localization in carbon nanofiber fabricated using electrospinning. J Appl Phys 94(3):1721–1727. doi:10.1063/1.1587268

29. Zhou YX, Freitag M, Hone J, Staii C, Johnson AT, Pinto NJ, MacDiarmid AG (2003) Fabrication and electrical characterization of polyaniline-based nanofibers with diameter below 30 nm. Appl Phys Lett 83(18):3800–3802. doi:10.1063/1.1622108

30. Wang Y, Ramos I, and Santiago-Aviles J (2007) Electrical characterization of a single electrospun porous SnO2 nanoribbon in ambient air. Nanotechnology 18 (43):435704 (1–4). doi:4357041O.1088/0957-4484/18/43/435704

31. Zhou ZP, Lai CL, Zhang LF, Qian Y, Hou HQ, Reneker DH, Fong H (2009) Development of carbon nanofibers from aligned electrospun polyacrylonitrile nanofiber bundles and characterization of their microstructural, electrical, and mechanical properties. Polymer 50(13):2999–3006. doi:10.1016/j.polymer.2009.04.058

32. Ferain I, Colinge CA, Colinge JP (2011) Multigate transistors as the future of classical metal-oxide-semiconductor field-effect transistors. Nature 479(7373):310–316. doi:10.1038/nature10676

33. Ionescu AM, Riel H (2011) Tunnel field-effect transistors as energy-efficient electronic switches. Nature 479(7373):329–337. doi:10.1038/nature10679

34. Kergoat L, Piro B, Berggren M, Horowitz G, Pham MC (2012) Advances in organic transistor-based biosensors: from organic electrochemical transistors to electrolyte-gated organic field-effect transistors. Anal Bioanal Chem 402(5):1813–1826. doi:10.1007/s00216-011-5363-y

35. Natali D, Caironi M (2012) Charge injection in solution-processed organic field-effect transistors: physics. Models Charact Methods. Adv Mater 24(11):1357–1387. doi:10.1002/adma.201104206

36. Sokolov AN, Tee BCK, Bettinger CJ, Tok JBH, Bao ZN (2012) Chemical and engineering approaches to enable organic field-effect transistors for electronic skin applications. Acc Chem Res 45(3):361–371. doi:10.1021/ar2001233

37. Wang CL, Dong HL, Hu WP, Liu YQ, Zhu DB (2012) Semiconducting pi-conjugated systems in field-effect transistors: a material odyssey of organic electronics. Chem Rev 112(4):2208–2267. doi:10.1021/cr100380z

38. Pinto NJ, Johnson AT, MacDiarmid AG, Mueller CH, Theofylaktos N, Robinson DC, Miranda FA (2003) Electrospun polyaniline/polyethylene oxide nanofiber field-effect transistor. Appl Phys Lett 83(20):4244–4246. doi:10.1063/1.1627484

39. Wu H, Lin DD, Pan W (2006) Fabrication, assembly, and electrical characterization of CuO nanofibers. Appl Phys Lett 89(13). doi:133125(1)-(3).10.1063/1.2355474

40. Wang W, Li ZY, Xu XR, Dong B, Zhang HN, Wang ZJ, Wang C, Baughman RH, Fang SL (2011) Au-Doped polyacrylonitrile-polyaniline core-shell electrospun nanofibers having high field-effect mobilities. Small 7(5):597–600. doi:10.1002/smll.201001716

41. Wang W, Lu XF, Li ZY, Lei JY, Liu XC, Wang ZJ, Zhang HN, Wang C (2011) one-dimensional polyelectrolyte/polymeric semiconductor core/shell structure: sulfonated poly(arylene ether ketone)/polyaniline nanofibers for organic field-effect transistors. Adv Mater 23(43):5109–5112. doi:10.1002/adma.201102125

42. Jaruwongrungsee K, Tuantranont A, Wanna Y, Wisitsoraat A, Lomas T, and Ieee (2007) Quartz crystal microbalance humidity sensor using electrospun PANI micro/nano dots (2007). In: 7th Ieee Conference on Nanotechnology, Vol 1. pp 3316–3319. doi:10.1109/nano.2007.4601198

43. Li ZY, Zhang HN, Zheng W, Wang W, Huang HM, Wang C, MacDiarmid AG, Wei Y (2008) Highly sensitive and stable humidity nanosensors based on LiCl doped TiO2 electrospun nanofibers. J Am Chem Soc 130(15):5036–5037. doi:10.1021/ja800176s

44. Corres JM, Garcia YR, Arregui FJ, Matias IR (2011) Optical fiber humidity sensors using PVdF electrospun nanowebs. IEEE Sens J 11(10):2383–2387. doi:10.1109/jsen.2011.2123881

45. Liu HQ, Kameoka J, Czaplewski DA, Craighead HG (2004) Polymeric nanowire chemical sensor. Nano Lett 4(4):671–675. doi:10.1021/nl049826f

46. Ding B, Kim JH, Miyazaki Y, Shiratori SM (2004) Electrospun nanofibrous membranes coated quartz crystal microbalance as gas sensor for NH3 detection. Sens Actuators B-Chem 101(3):373–380. doi:10.1016/j.snb.2004.04.008

47. Kawasaki H, Ueda T, Suda Y, Ohshima T (2004) Properties of metal doped tungsten oxide thin films for NOx gas sensors grown by PLD method combined with sputtering process. Sens Actuators B-Chem 100(1–2):266–269. doi:10.1016/j.snb.2003.12.052

48. Hao R, Yuan JY, Peng Q (2006) Fabrication and sensing behavior of Cr2O3 nanofibers via in situ gelation and electrospinning. Chem Lett 35(11):1248–1249. doi:10.1246/cl.2006.1248

49. Kim ID, Rothschild A, Lee BH, Kim DY, Jo SM, Tuller HL (2006) Ultrasensitive chemiresistors based on electrospun TiO2 nanofibers. Nano Lett 6(9):2009–2013. doi:10.1021/nl061197h

50. Zhang X, Ji L, Toprakci O, Liang Y, Alcoutlabi M (2011) electrospun nanofiber-based anodes, cathodes, and separators for advanced Lithium–Ion batteries. Polym Rev 51(3):239–264. doi:10.1080/15583724.2011.593390

51. Guo Y-G, Hu J-S, Wan L-J (2008) nanostructured materials for electrochemical energy conversion and storage devices. Adv Mater 20(23):4384–4384 (vol 20, p 2878)

52. Kim C, Yang KS, Kojima M, Yoshida K, Kim YJ, Kim YA, Endo M (2006) Fabrication of electrospinning-derived carbon nanofiber webs for the anode material of lithium-ion secondary batteries. Adv Funct Mater 16(18):2393–2397. doi:10.1002/adfm.200500911

53. Wang L, Yu Y, Chen PC, Zhang DW, Chen CH (2008) Electrospinning synthesis of C/Fe3O4 composite nanofibers and their application for high performance lithium-ion batteries. J Power Sources 183(2):717–723. doi:10.1016/j.jpowsour.2008.05.079

54. Wang L, Yu Y, Chen P-C, Chen C-H (2008) Electrospun carbon–cobalt composite nanofiber as an anode material for lithium ion batteries. Scripta Mater 58(5):405–408. doi:10.1016/j.scriptamat.2007.10.024

55. Fan Q, Whittingham MS (2007) Electrospun manganese oxide nanofibers as anodes for lithium-ion batteries. Electrochem Solid State Lett 10(3):A48–A51. doi:10.1149/1.2422749

56. Ding YH, Zhang P, Long ZL, Jiang Y, Huang JN, Yan WJ, Liu G (2008) Synthesis and electrochemical properties of Co3O4 nanofibers as anode materials for lithium-ion batteries. Mater Lett 62(19):3410–3412. doi:10.1016/j.matlet.2008.03.033

57. Gu YX, Jian FF, Wang X (2008) Synthesis and characterization of nanostructured Co3O4 fibers used as anode materials for lithium ion batteries. Thin Solid Films 517(2):652–655. doi:10.1016/j.tsf.2008.07.026

58. Ban CM, Chernova NA, Whittingham MS (2009) Electrospun nano-vanadium pentoxide cathode. Electrochem Commun 11(3):522–525. doi:10.1016/j.elecom.2008.11.051

59. Ji LW, Jung KH, Medford AJ, Zhang XW (2009) Electrospun polyacrylonitrile fibers with dispersed Si nanoparticles and their electrochemical behaviors after carbonization. J Mater Chem 19(28):4992–4997. doi:10.1039/b903165k

60. Ji LW, Lin Z, Medford AJ, Zhang XW (2009) In-situ encapsulation of nickel particles in electrospun carbon nanofibers and the resultant electrochemical performance. Chem-a Eur J 15(41):10718–10722. doi:10.1002/chem.200902012

61. Ji LW, Lin Z, Medford AJ, Zhang XW (2009) Porous carbon nanofibers from electrospun polyacrylonitrile/SiO2 composites as an energy storage material. Carbon 47(14):3346–3354. doi:10.1016/j.carbon.2009.08.002

62. Ji LW, Medford AJ, Zhang XW (2009) Porous carbon nanofibers loaded with manganese oxide particles: formation mechanism and electrochemical performance as energy-storage materials. J Mater Chem 19(31):5593–5601. doi:10.1039/b905755b

63. Ji LW, Zhang XW (2009) Manganese oxide nanoparticle-loaded porous carbon nanofibers as anode materials for high-performance lithium-ion batteries. Electrochem Commun 11(4):795–798. doi:10.1016/j.elecom.2009.01.039

64. Ji LW, Zhang XW (2009) Fabrication of porous carbon/Si composite nanofibers as high-capacity battery electrodes. Electrochem Commun 11(6):1146–1149. doi:10.1016/j.elecom.2009.03.042

65. Gu YX, Chen DR, Jiao ML (2005) Synthesis and electrochemical properties of nanostructured LiCoO2 fibers as cathode materials for lithium-ion batteries. J Phys Chem B 109(38):17901–17906. doi:10.1021/jp0521813
66. Gu YX, Chen DR, Jiao XL, Liu FF (2007) LiCoO2-MgO coaxial fibers: co-electrospun fabrication, characterization and electrochemical properties. J Mater Chem 17(18):1769–1776. doi:10.1039/b614205b
67. Toprakci O, Ji LW, Lin Z, Toprakci HAK, Zhang XW (2011) Fabrication and electrochemical characteristics of electrospun LiFePO4/carbon composite fibers for lithium-ion batteries. J Power Sources 196(18):7692–7699. doi:10.1016/j.jpowsour.2011.04.031
68. Ding YH, Zhang P, Long ZL, Jiang Y, Xu F (2009) Morphology and electrochemical properties of Al doped LiNi1/3Co1/3Mn1/3O2 nanofibers prepared by electrospinning. J Alloy Compd 487(1–2):507–510. doi:10.1016/j.jallcom.2009.08.002
69. Gao K, Hu X, Dai C, Yi T (2006) Crystal structures of electrospun PVDF membranes and its separator application for rechargeable lithium metal cells. Mater Sci Eng, B 131(1–3):100–105. doi:10.1016/j.mseb.2006.03.035
70. Cho TH, Sakai T, Tanase S, Kimura K, Kondo Y, Tarao T, Tanaka M (2007) Electrochemical performances of polyacrylonitrile nanofiber-based nonwoven separator for lithium-ion battery. Electrochem Solid State Lett 10(7):A159–A162. doi:10.1149/1.2730727
71. Ding YH, Di W, Jiang Y, Xu F, Long ZL, Ren FM, Zhang P (2009) The morphological evolution, mechanical properties and ionic conductivities of electrospinning P(VDF-HFP) membranes at various temperatures. Lonics 15(6):731–734. doi:10.1007/s11581-009-0326-4
72. Ding YH, Zhang P, Long ZL, Jiang Y, Xu F, Di W (2009) The ionic conductivity and mechanical property of electrospun P(VdF-HFP)/PMMA membranes for lithium ion batteries. J Membr Sci 329(1–2):56–59. doi:10.1016/j.memsci.2008.12.024
73. Liang YZ, Ji LW, Guo BK, Lin Z, Yao YF, Li Y, Alcoutlabi M, Qiu YP, Zhang XW (2011) Preparation and electrochemical characterization of ionic-conducting lithium lanthanum titanate oxide/polyacrylonitrile submicron composite fiber-based lithium-ion battery separators. J Power Sources 196(1):436–441. doi:10.1016/j.jpowsour.2010.06.088
74. Bajon R, Balaji S, and Guo SM (2009) Electrospun nafion nanofiber for proton exchange membrane fuel cell application. J Fuel Cell Sci Technol 6(3):031004(1–6) doi:10.1115/1.3005577
75. Jang WG, Hou J, Byun HS (2011) Preparation and characterization of PVdF nanofiber ion exchange membrane for the PEMFC application. Desalin Water Treat 34(1–3):315–320. doi:10.5004/dwt.2011.2871
76. Molla S, Compan V (2011) Polyvinyl alcohol nanofiber reinforced Nafion membranes for fuel cell applications. J Membr Sci 372(1–2):191–200. doi:10.1016/j.memsci.2011.02.001
77. Molla S, Compan V, Gimenez E, Blazquez A, Urdanpilleta I (2011) Novel ultrathin composite membranes of Nafion/PVA for PEMFCs. Int J Hydrogen Energy 36(16):9886–9895. doi:10.1016/j.ijhydene.2011.05.074
78. Tamura T, Kawakami H (2010) Aligned electrospun nanofiber composite membranes for fuel cell electrolytes. Nano Lett 10(4):1324–1328. doi:10.1021/nl1007079
79. Liu W, Wang S, Xiao M, Han D, Meng Y (2012) A proton exchange membrane fabricated from a chemically heterogeneous nonwoven with sandwich structure by the program-controlled co-electrospinning process. Chem Commun 48(28):3415–3417. doi:10.1039/c2cc16952e
80. Kenawy ER, Bowlin GL, Mansfield K, Layman J, Simpson DG, Sanders EH, Wnek GE (2002) Release of tetracycline hydrochloride from electrospun poly(ethylene-co-vinylacetate), poly(lactic acid), and a blend. J Controlled Release 81(1–2):57–64. doi:10.1016/s0168-3659(02)00041-x
81. Kim K, Luu YK, Chang C, Fang DF, Hsiao BS, Chu B, Hadjiargyrou M (2004) Incorporation and controlled release of a hydrophilic antibiotic using poly (lactide-co-glycolide)-based electrospun nanofibrous scaffolds. J Controlled Release 98(1):47–56. doi:10.1016/j.jconrel.2004.04.009

82. Feng K, Sun H, Bradley MA, Dupler EJ, Giannobile WV, Ma PX (2010) Novel antibacterial nanofibrous PLLA scaffolds. J Controlled Release 146(3):363–369. doi:10.1016/j.jconrel.2010.05.035

83. Xu XL, Yang LX, Xu XY, Wang X, Chen XS, Liang QZ, Zeng J, Jing XB (2005) Ultrafine medicated fibers electrospun from W/O emulsions. J Controlled Release 108(1):33–42. doi:10.1016/j.jconrel.2005.07.021

84. Xu X, Chen X, Xu X, Lu T, Wang X, Yang L, Jing X (2006) BCNU-loaded PEG–PLLA ultrafine fibers and their in vitro antitumor activity against Glioma C6 cells. J Controlled Release 114(3):307–316. doi:10.1016/j.jconrel.2006.05.031

85. Xie J, Wang C-H (2006) Electrospun micro- and nanofibers for sustained delivery of paclitaxel to treat C6 glioma in vitro. Pharm Res 23(8):1817–1826. doi:10.1007/s11095-006-9036-z

86. Gao H, Gu Y, Ping Q (2007) The implantable 5-fluorouracil-loaded poly (L-lactic acid) fibers prepared by wet-spinning from suspension. J Controlled Release 118(3):325–332. doi:10.1016/j.jconrel.2006.12.028

87. Hartman O, Zhang C, Adams EL, Farach-Carson MC, Petrelli NJ, Chase BD, Rabolt JF (2009) Microfabricated electrospun collagen membranes for 3-D cancer models and drug screening applications. Biomacromolecules 10(8):2019–2032. doi:10.1021/bm8012764

88. Ignatova MG, Manolova NE, Toshkova RA, Rashkov IB, Gardeva EG, Yossifova LS, Alexandrov MT (2010) Electrospun nanofibrous mats containing quaternized Chitosan and Polylactide with in vitro antitumor activity against HeLa cells. Biomacromolecules 11(6):1633–1645. doi:10.1021/bm100285n

89. Kim Y-J, Bae H-I, Kwon OK, Choi M-S (2009) Three-dimensional gastric cancer cell culture using nanofiber scaffold for chemosensitivity test. Int J Biol Macromol 45(1):65–71. doi:10.1016/j.ijbiomac.2009.04.003

90. Kumbar SG, Nair LS, Bhattacharyya S, Laurencin CT (2006) Polymeric nanofibers as novel carriers for the delivery of therapeutic molecules. J Nanosci Nanotechnol 6(9–10):2591–2607. doi:10.1166/jnn.2006.462

91. Sill TJ, von Recum HA (2008) Electro spinning: applications in drug delivery and tissue engineering. Biomaterials 29(13):1989–2006. doi:10.1016/j.biomaterials.2008.01.011

92. Venugopal J, Prabhakaran MP, Low S, Choon AT, Deepika G, Dev VRG, Ramakrishna S (2009) Continuous nanostructures for the controlled release of drugs. Curr Pharm Des 15(15):1799–1808

93. Park Y, Kang E, Kwon O-J, Hwang T, Park H, Lee JM, Kim JH, Yun C-O (2010) Ionically crosslinked Ad/chitosan nanocomplexes processed by electrospinning for targeted cancer gene therapy. J Controlled Release 148(1):75–82. doi:10.1016/j.jconrel.2010.06.027

94. Zeng J, Xu X, Chen X, Liang Q, Bian X, Yang L, Jing X (2003) Biodegradable electrospun fibers for drug delivery. J Controlled Release 92(3):227–231. doi:10.1016/s0168-3659(03)00372-9

95. Zeng J, Yang L, Liang Q, Zhang X, Guan H, Xu X, Chen X, Jing X (2005) Influence of the drug compatibility with polymer solution on the release kinetics of electrospun fiber formulation. J Controlled Release 105(1–2):43–51. doi:10.1016/j.jconrel.2005.02.024

96. Liang DH, Luu YK, Kim KS, Hsiao BS, Hadjiargyrou M, Chu B (2005) In vitro non-viral gene delivery with nanofibrous scaffolds. Nucleic Acids Res 33(19). doi:E17010.1093/nar/gnil171

97. Liao IC, Chen S, Liu JB, Leong KW (2009) Sustained viral gene delivery through core-shell fibers. J Controlled Release 139(1):48–55. doi:10.1016/j.jconrel.2009.06.007

98. Saraf A, Baggett LS, Raphael RM, Kasper FK, Mikos AG (2010) Regulated non-viral gene delivery from coaxial electrospun fiber mesh scaffolds. J Controlled Release 143(1):95–103. doi:10.1016/j.jconrel.2009.12.009

99. Langer R, Vacanti J (1993) Tissue engineering. Science 260(5110):920–926. doi:10.1126/science.8493529

100. Rosenberg MD (1963) Cell guidance by alterations in monomolecular films. Science 139(3553):411. doi:10.1126/science.139.3553.411

101. Ma ZW, Kotaki M, Inai R, Ramakrishna S (2005) Potential of nanofiber matrix as tissue-engineering scaffolds. Tissue Eng 11(1–2):101–109. doi:10.1089/ten.2005.11.101
102. Murphy WL, Peters MC, Kohn DH, Mooney DJ (2000) Sustained release of vascular endothelial growth factor from mineralized poly (lactide-co-glycolide) scaffolds for tissue engineering. Biomaterials 21(24):2521–2527. doi:10.1016/s0142-9612(00)00120-4
103. Moioli EK, Clark PA, Xin X, Lal S, Mao JJ (2007) Matrices and scaffolds for drug delivery in dental, oral and craniofacial tissue engineering. Adv Drug Deliv Rev 59(4–5):308–324. doi:10.1016/j.addr.2007.03.019
104. Martins A, Duarte ARC, Faria S, Marques AP, Reis RL, Neves NM (2010) Osteogenic induction of hBMSCs by electrospun scaffolds with dexamethasone release functionality. Biomaterials 31(22):5875–5885. doi:10.1016/j.biomaterials.2010.04.010
105. Wang F, Li Z, Khan M, Tamama K, Kuppusamy P, Wagner WR, Sen CK, Guan J (2010) Injectable, rapid gelling and highly flexible hydrogel composites as growth factor and cell carriers. Acta Biomater 6(6):1978–1991. doi:10.1016/j.actbio.2009.12.011
106. Loh XJ, Peh P, Liao S, Sng C, Li J (2010) Controlled drug release from biodegradable thermoresponsive physical hydrogel nanofibers. J Controlled Release 143(2):175–182. doi:10.1016/j.jconrel.2009.12.030
107. Kolambkar YM, Dupont KM, Boerckel JD, Huebsch N, Mooney DJ, Hutmacher DW, Guldberg RE (2011) An alginate-based hybrid system for growth factor delivery in the functional repair of large bone defects. Biomaterials 32(1):65–74. doi:10.1016/j.biomaterials.2010.08.074
108. Nair LS, Bhattacharyya S, Laurencin CT (2004) Development of novel tissue engineering scaffolds via electrospinning. Expert Opin Biol Ther 4(5):659–668. doi:10.1517/eobt.4.5.659.31057
109. Chiu JB, Luu YK, Fang DF, Hsiao BS, Chu B, Hadjiargyrou M (2005) Electrospun nanofibrous scaffolds for biomedical applications. J Biomed Nanotechnol 1(2):115–132. doi:10.1166/jbn.2005.018
110. Pham QP, Sharma U, Mikos AG (2006) Electrospinning of polymeric nanofibers for tissue engineering applications: a review. Tissue Eng 12(5):1197–1211. doi:10.1089/ten.2006.12.1197
111. Barnes CP, Sell SA, Boland ED, Simpson DG, Bowlin GL (2007) Nanofiber technology: designing the next generation of tissue engineering scaffolds. Adv Drug Deliv Rev 59(14):1413–1433. doi:10.1016/j.addr.2007.04.022
112. Liang D, Hsiao BS, Chu B (2007) Functional electrospun nanofibrous scaffolds for biomedical applications. Adv Drug Deliv Rev 59(14):1392–1412. doi:10.1016/j.addr.2007.04.021
113. Martins A, Araujo JV, Reis RL, Neves NM (2007) Electrospun nanostructured scaffolds for tissue engineering applications. Nanomedicine 2(6):929–942. doi:10.2217/17435889.2.6.929
114. Murugan R, Ramakrishna S (2007) Design strategies of tissue engineering scaffolds with controlled fiber orientation. Tissue Eng 13(8):1845–1866. doi:10.1089/ten.2006.0078
115. Sell S, Barnes C, Smith M, McClure M, Madurantakam P, Grant J, McManus M, Bowlin G (2007) Extracellular matrix regenerated: tissue engineering via electrospun biomimetic nanofibers. Polym Int 56(11):1349–1360. doi:10.1002/pi.2344
116. Zhang YZ, Su B, Venugopal J, Ramakrishna S, Lim CT (2007) Biomimetic and bioactive nanofibrous scaffolds from electrospun composite nanofibers. Int J Nanomed 2(4):623–638
117. Agarwal S, Wendorff JH, Greiner A (2008) Use of electrospinning technique for biomedical applications. Polymer 49(26):5603–5621. doi:10.1016/j.polymer.2008.09.014
118. Ashammakhi N, Ndreu A, Nikkola L, Wimpenny I, Yang Y (2008) Advancing tissue engineering by using electrospun nanofibers. Regenerative Medicine 3(4):547–574. doi:10.2217/17460751.3.4.547
119. Ayres CE, Jha BS, Meredith H, Bowman JR, Bowlin GL, Henderson SC, Simpson DG (2008) Measuring fiber alignment in electrospun scaffolds: a user's guide to the 2D fast Fourier transform approach. J Biomaterials Science-Polymer Edition 19(5):603–621. doi:10.1163/156856208784089643

120. Fang J, Niu HT, Lin T, Wang XG (2008) Applications of electrospun nanofibers. Chin Sci Bull 53(15):2265–2286. doi:10.1007/s11434-008-0319-0
121. Gelain F (2008) Novel opportunities and challenges offered by nanobiomaterials in tissue engineering. Int J Nanomed 3(4):415–424
122. Kumbar SG, James R, Nukavarapu SP, Laurencin CT (2008) Electrospun nanofiber scaffolds: engineering soft tissues. Biomed Mater 3(3):034002. doi:10.1088/1748-6041/3/3/034002
123. Martins A, Reis RL, Neves NM (2008) Electrospinning: processing technique for tissue engineering scaffolding. Int Mater Rev 53(5):257–274. doi:10.1179/174328008x353547
124. Venugopal J, Low S, Choon AT, Ramakrishna S (2008) Interaction of cells and nanofiber scaffolds in tissue engineering. Journal of Biomedical Materials Research Part B-Applied Biomaterials 84B(1):34–48. doi:10.1002/jbm.b.30841
125. Burger C, Hsiao BS, and Chu B (2006) Nanofibrous materials and their applications. In: Annual review of materials research, vol 36. Annual Review of Materials Research. Annual Reviews, USA, pp 333–368. doi:10.1146/annurev.matsci.36.011205.123537
126. Bosworth L, and Downes S (2009) Biocompatible three-dimensional scaffolds for tendon tissue engineering using electrospinning. cellular response to biomaterials. In: Disilvio (ed) Woodhead Publishing in Materials, WOODHEAD PUBL LTD, England, pp 3–27. doi:10.1533/9781845695477.1.3
127. Focarete ML, Gualandi C, Moroni L (2009) Working with electrospun scaffolds: some practical hints for tissue engineers. In: Moroni L (ed) Electrospun Nanofibers Research: Recent Developments. nanotechnology science and technology. Nova Science Publishers, USA, pp 19–34
128. Haghi AK (2009) Electrospun biodegradable and biocompatible natural nanofibers: a detailed review. In: Highi AK (ed) Electrospun nanofibers research: recent developments, nanotechnology science and technology. Nova Science Publishers, USA, pp 171–205
129. Park K, Park WH, Son JS, and Han DK (2009) Biomedical polymer nanofibers for emerging technology. In: Shi D (ed) nanoscience in biomedicine. Springer, Germany, pp 21–42. doi:10.1007/978-3-540-49661-8_2
130. Powell HM, Boyce ST (2009) Cultured skin substitutes. In: Orgill DP (ed) Biomaterials for Treating Skin Loss. Woodhead Publishing in Materials, Woodhead Publ Ltd, England, p 183–206. doi:10.1533/9781845695545.3.183
131. Venugopal J, Prabhakaran MP, Zhang YZ, Deepika G, Dev VRG, Low S, Choon AT, Ramakrishna S (2009) nanotechnology: a global challenge in healthcare. In: Haghi AK (ed) Electrospun nanofibers research: recent developments. nanotechnology science and technology. Nova Science Publishers, USA, pp 253–277
132. Wise SG, Mithieux SM, and Weiss AS (2009) Engineered tropoelastin and elastin-based biomaterials. In: McPherson A (ed) Advances in protein chemistry and structural biology, vol 78. Elsevier, USA, pp 1–24. doi:10.1016/s1876-1623(09)78001-5
133. Bajgai MP, Aryal S, Kim HY (2010) Smart polymeric electrospun nanofiber for biological applications. In: Li S (ed) Smart Polymer Materials for biomedical applications. materials science and technologies. Nova Science Publishers, USA, pp 119–135
134. Selim KMK, Kang IK (2010) Preparation of polymer composite nanofibers by electrospinning and their biomedical application. In: Lechkov M (ed) Encyclopedia of polymer composites: properties, performance and applications, polymer science and technology series. Nova Science Publishers, USA, pp 755–806
135. Teo WE, Kaur S, Ramakrishna S (2010) Electrospun polymer nanocomposite fibers: fabrication and physical properties. In: Tjong SC (ed) Physical properties and applications of polymer nanocomposites. Woodhead, England, pp 616–637
136. Chahine NO, Chao PHG (2011) Micro and nanotechnologies for tissue engineering. In: Burdick JA (ed) Biomaterials for tissue engineering applications: a review of the past and future trends. Springer, US, pp 139–178. doi:10.1007/978-3-7091-0385-2_6

137. Haghi AK (2011) Practical hints on electrospinning of polymeric nanosized fibers. In: Haghi AK (ed) Modern concepts in nanotechnology research. nanotechnology science and technology. Nova Science Publishers, USA, pp 61–76

138. James R, Toti US, Laurencin CT, Kumbar SG (2011) Electrospun nanofibrous scaffolds for engineering soft connective tissues. In: Hurst SJ (ed) Biomedical nanotechnology: methods and protocols, vol 726, Methods in Molecular BiologyHumana Press, USA, pp 243–258

139. Rehman IU, Khan AS (2011) Dental regeneration. In: Bosworth LA (ed) Electrospinning for tissue regeneration. Woodhead, England, pp 280–297

140. Simonet M, Driessen-Mol A, Baaijens FPT, Bouten CVC (2011) Heart valve tissue regeneration. In: Bosworth LA (ed) Electrospinning for tissue regeneration. Woodhead, England, pp 202–224

141. Subramanian A, Krishnan UM, Sethuraman S (2011) Skin tissue regeneration. Electrospinning for tissue regeneration. Woodhead, England, pp 298–316

142. Van der Schueren L, De Clerck K (2011) Nanofibrous textiles in medical applications. Handbook of Medical Textiles, vol 100. Woodhead, England, pp 547–566

143. Wan JDD, Downes S, Dunne MJ, Cosgrove KE (2011) Cell culture systems for pancreatic research. Electrospinning for tissue regeneration. Woodhead, England, pp 359–371

144. Wang C, Koh H, Ramakrishna S, Liao S (2011) Nerve tissue regeneration. Electrospinning for tissue regeneration. Woodhead, England, pp 168–201

145. Wang L, Ryan AJ (2011) Introduction to electrospinning. Electrospinning for tissue regeneration. Woodhead, England, pp 3–33

146. Jayakumar R, Prabaharan M, Shalumon KT, Chennazhi KP, and Nair SV (2012) Biomedical applications of Polymer/Silver composite nanofibers. In: Jayakumar R, Nair SV (eds) Biomedical applications of polymeric nanofibers, vol 246. Advances in Polymer Science. Springer, Gemany, pp 263–82. doi:10.1007/12_2011_123

147. Katsanevakis E, Wen XJ, Zhang N (2012) Creating electrospun nanofiber-based biomimetic scaffolds for bone regeneration. In: Jayakumar R, Nair SV (eds) Biomedical applications of polymeric nanofibers, vol 246. Advances in Polymer Science. Springer, Gemany, pp 63–100. doi:10.1007/12_2011_131

148. Prabaharan M, Jayakumar R, and Nair SV (2012) Electrospun nanofibrous scaffolds-current status and prospects in drug delivery. In: Jayakumar R, Nair SV (eds) Biomedical applications of polymeric nanofibers, vol 246. Advances in Polymer Science. Springer, Gemany, pp 241–262. doi:10.1007/12_2011_125

149. Supaphol P, Suwantong O, Sangsanoh P, Srinivasan S, Jayakumar R, and Nair SV (2012) Electrospinning of biocompatible polymers and their potentials in biomedical applications. In: Jayakumar R, Nair SV (eds) Biomedical applications of polymeric nanofibers, vol 246. Advances in Polymer Science. Springer, Gemany, pp 213–239, 241–262. doi:10.1007/12_2011_143

150. Bognitzki M, Hou H, Ishaque M, Frese T, Hellwig M, Schwarte C, Schaper A, Wendorff JH, Greiner A (2000) Polymer, Metal, and hybrid nano- and mesotubes by coating degradable polymer template fibers (TUFT process). Adv Mater 12(9):637–640. doi:10.1002/(sici)1521-4095(200005)12:9<637:aid-adma637>3.0.co;2-w

151. Li D, Xia YN (2003) Fabrication of Titania nanofibers by electrospinning. Nano Lett 3(4):555–560. doi:10.1021/nl034039o

152. Ki CS, Gang EH, Um IC, Park YH (2007) Nanofibrous membrane of wool keratose/silk fibroin blend for heavy metal ion adsorption. J Membr Sci 302(1–2):20–26. doi:10.1016/j.memsci.2007.06.003

153. Vu D, Li ZY, Zhang HN, Wang W, Wang ZJ, Xu XR, Dong B, Wang C (2012) Adsorption of Cu (II) from aqueous solution by anatase mesoporous TiO2 nanofibers prepared via electrospinning. J Colloid Interface Sci 367:429–435. doi:10.1016/j.jcis.2011.09.088

154. Kaur S, Kotaki M, Ma Z, Gopal R, Ramakrishna S, Ng SC (2006) Oligosaccharide functionalized nanofibrous membrane. Int J Nanosci 05(01):1–11. doi:10.1142/S0219581X06004206

155. Zhang X, Du AJ, Lee P, Sun DD, Leckie JO (2008) TiO2 nanowire membrane for concurrent filtration and photocatalytic oxidation of humic acid in water. J Membr Sci 313(1–2):44–51. doi:10.1016/j.memsci.2007.12.045

156. Moradzadegan A, Ranaei-Siadat SO, Ebrahim-Habibi A, Barshan-Tashnizi M, Jalili R, Torabi SF, Khajeh K (2010) Immobilization of acetylcholinesterase in nanofibrous PVA/BSA membranes by electrospinning. Eng Life Sci 10(1):57–64. doi:10.1002/elsc.200900001

157. Piperno S, Tse Sum Bui B, Haupt K, Gheber LA (2011) Immobilization of molecularly imprinted polymer nanoparticles in electrospun poly (vinyl alcohol) nanofibers. Langmuir 27(5):1547–1550. doi:10.1021/la1041234

158. Kim WJ, Chang JY (2011) Molecularly imprinted polyimide nanofibers prepared by electrospinning. Mater Lett 65(9):1388–1391. doi:10.1016/j.matlet.2011.02.010

159. Wang J, Yao HB, He DA, Zhang CL, Yu SH (2012) Facile fabrication of gold nanoparticles-poly(vinyl alcohol) electrospun water-stable nanofibrous mats: efficient substrate materials for biosensors. ACS Appl Mat Interfaces 4(4):1963–1971. doi:10.1021/am300391j

Author Biography

A/Prof. Zhenyu Li studied in China and received his degree in Polymer Chemistry and Physics from the Jilin University in 2007. From 2009 to 2010, he was employed by the College of Engineering in the University of Nebraska-Lincoln. From 2011, he was appointed Associated Professor in Jilin University. His research interests are 1D nanostructures, chemiresistors, field-effect transistors, and photocatalysis.

Prof. Ce Wang studied in, USA, Austria, and China. She received her degree in Organic polymer materials from the Vienna Technical University in 1993. From 1997, she was employed by the Jilin University by the Department of Chemistry as A/Professor. In 1999, she was appointed full professor at the Jilin University. Her research interests are organic/inorganic hybrid materials, organic functional nanomaterials, multi-phase composite nanomaterials, and 1D electrospun nanostructures. Prof. Ce Wang has authored or co-authored more than 200 scientific papers published in the worldwide famous journals and 14 patents.

Z. Li and C. Wang, *One-Dimensional Nanostructures*, SpringerBriefs in Materials, DOI: 10.1007/978-3-642-36427-3, © The Author(s) 2013